CURRENT COMMUNICATIONS 3
In Cell & Molecular Biology

Apoptosis: The Molecular Basis of Cell Death

Edited by
L. David Tomei
The Ohio State University

Frederick O. Cope
Ross Laboratories

CSHL PRESS

Cold Spring Harbor Laboratory Press 1991

CURRENT COMMUNICATIONS 3
In Cell & Molecular Biology
Apoptosis: The Molecular Basis of Cell Death

Front Cover: Apoptosis of murine NS-1 cell occurring spontaneously in culture. Note the discrete nuclear fragments with characteristic segregation of compacted chromatin, the crowding of organelles, and the marked convolution of the cellular surface. (Courtesy of J.F.R. Kerr and B.V. Harmon. See related article by Kerr and Harmon, this volume.)

Back cover: Scanning electron micrograph of a C3H-10T1/2 mouse embryonic fibroblast which is undergoing apoptosis following serum deprivation. Note the loss of cell shape but maintenance of adhesion sites. (Courtesy of S.W. Hui and C.E. Wenner, Roswell Park Cancer Institute, and L.D. Tomei, The Ohio State University.)

Library of Congress Cataloging-in-Publication Data

Apoptosis : the molecular basis of cell death / edited by L. David
    Tomei, Frederick O. Cope.
        p. cm. — (Current communications in cell & molecular biology
    ; 3)
    Includes index.
    ISBN 0-87969-366-5
    1. Apoptosis. 2. Cell death. I. Tomei, L. David. II. Cope,
Frederick O. III. Series: Current communications in cell and
molecular biology ; 3.
QH671.A66   1991
574.87'65--dc20   91-9068
                                                   CIP

The articles published in this book have not been peer-reviewed. They express their authors' views, which are not necessarily endorsed by Cold Spring Harbor Laboratory.

All Cold Spring Harbor Laboratory Press publications may be ordered directly from Cold Spring Harbor Laboratory Press, 10 Skyline Drive, Plainview, New York 11803. Phone: 1-800-843-4388. In New York (516) 349-1930. FAX: (516) 349-1946.

Apoptosis:
The Molecular Basis
of Cell Death

SERIES EDITORS
John Inglis and Jan A. Witkowski
Cold Spring Harbor Laboratory

CURRENT COMMUNICATIONS
In Cell & Molecular Biology

Contents

Preface

The purpose of this book is to provide a glimpse into the dynamic process of scientific concept formation by presenting the thoughts of some of the scientists who are engaged in development of a new subject of inquiry, often from very different perspectives. Apoptosis is a concept that may eventually be considered as one of the most formidable in biology. However, at this time it is the subject of intense scrutiny, critical assessment, speculation, and skepticism from a remarkably wide range of fields within biology and medicine. Since this is the case, many of us feel that we have a ringside seat to watch an extraordinary process, which is the life of scientific inquiry.

This book has been written for the graduate student in the basic biological sciences and medicine who is interested in the newest concepts, as well as for the established researcher who may be seeking the conceptual framework that will bring together years of empirical observations. The authors believe that apoptosis is a seminal concept whose significance is equivalent to that of the cell cycle. Chapters are devoted to cell biology, gene expression, the immune system, radiobiology, cancer and anticancer therapy, developmental biology, and the perspective of evolutionary development. Regardless of whether the ideas expressed here are ever found to be correct or incorrect, if this book stimulates discussion and new scientific inquiry, then it will have been successful.

We thank John Inglis and Jan Witkowski, series editors, for their part in bringing about the meeting. We are grateful to the corporate sponsors and other funders for financial support and to the Banbury staff, who created an environment conducive to a productive meeting. We acknowledge the work of the staff of Cold Spring Harbor Laboratory Press, including Patricia Barker, technical editor, and Inez Sialiano, editorial assistant.

L.D.T.
F.O.C.

Special Support

The meeting at the Banbury Center on which this book is based was supported by funding from:

Abbott Laboratories
CNS Research
Metropolitan Life Foundation
The Ohio State University
Ross Laboratories

Corporate Sponsors

The meetings' program at Cold Spring Harbor Laboratory is supported by:

Alafi Capital Company
American Cyanamid Company
AMGen Inc.
Applied Biosystems, Inc.
Becton Dickinson and Company
Boehringer Mannheim Corporation
Bristol-Myers Squibb Company
Ciba-Geigy Corporation/Ciba-Geigy Limited
Diagnostic Products Corporation
E.I. du Pont de Nemours & Company
Eastman Kodak Company
Genentech, Inc.
Genetics Institute
Hoffmann-La Roche Inc.
Johnson & Johnson
Kyowa Hakko Kogyo Co., Ltd.
Life Technologies, Inc.
Eli Lilly and Company
Millipore Corporation
Monsanto Company
Pall Corporation
Perkin-Elmer Cetus Instruments
Pfizer Inc.
Pharmacia Inc.
Schering-Plough Corporation
SmithKline Beecham Pharmaceuticals
The Wellcome Research Laboratories,
    Burroughs Wellcome Co.
Wyeth-Ayerst Research

Apoptosis:
The Molecular Basis
of Cell Death

Introduction

L. David Tomei[1] **and Frederick O. Cope**[2]
[1]The Ohio State University
Comprehensive Cancer Center
Columbus, Ohio 43210
[2]Ross Laboratories
Columbus, Ohio 43215-1754

Much of the recent literature on apoptosis cites the landmark paper of Kerr, Wyllie, and Currie published in 1972. Scientists obviously do not work in an intellectual and scientific vacuum, and the basic concept of deliberate cell death was based on several decades of observations made by numerous investigators who contributed to the literature. However, the quality that made the paper by Kerr et al. important to science was that it was arguably the first time cell death had been integrated into an empirically verifiable hypothesis. These authors extended their writing beyond descriptive phrases of cellular death and gave recognition to the fact that the orderly death of cells in tissues, organs, and developing embryos was as fundamentally important to the understanding of biology as was mitosis. Until then, the fact that normal tissue turnover involved continuous cell death remained a tacit assumption.

A question that must be important to all who study science and how science works is how did it come about that, through much of the next two decades, the term *apoptosis* came into scattered use as a descriptive term in cytology and pathology, while the importance of the hypothesis remained elusive? The major reason that apoptosis has not been fully integrated into cell and molecular biology may be that cell biologists assume that toxic insult is the direct cause of death, and if we understand the molecular nature of the insult, we understand why cells die. Although this view is accurate in many instances, we now know that it is generally not true. However, this fact is not intuitively evident if the nature of the investigation is the mechanism of death from exposure to chemicals or radiation, since these agents are, in fact, capable of severe cellular dam-

Apoptosis: The Molecular Basis of Cell Death
Copyright 1991 Cold Spring Harbor Laboratory Press 0-87969-366-5/91 $3.00 + 00

age. Alternatively, death in normal tissues and organs during development or involution may be intuitively associated with "deliberate" processes, yet searches for mechanisms conventionally are focused on the assumption that if a cell dies, it must have been the target of an externally initiated catastrophic event. However, numerous studies published during the last few years lend growing support to the view that cells can survive what may have been once considered catastrophic damage by cytotoxic chemicals or ionizing radiation without interference with the nature of the primary damage or any interference with normal repair of that damage. It is becoming increasingly evident that cell death under a variety of conditions and environments is the product of a precise biochemical cascade. Therefore, perhaps the greatest value of the concept of apoptosis lies in the inherent value of scientific hypothesis; apoptosis as an hypothesis provides a new conceptual framework whereby we explain phenomena *and* successfully predict new phenomena.

Cell death is no longer considered only in terms of catastrophic failure of cell integrity produced by molecular damage. Rather, it is now important to consider whether death is an endpoint of a cascade of common metabolic events. On the basis of the general concept of apoptosis, molecular damage directly resulting from toxic insult should not be presumed to be the cause of death. Conversely, survival of cells should not be attributed to a lack of molecular damage without further analyses. Instead, molecular damage must now be considered as a possible signal in a pathway that converges on a limited number of cellular changes that immediately precede death in some, but notably not all, cells that sustain damage. In this view, these hypothetical final changes cause the catastrophic failure recognized as cell death instead of the original damage introduced by the toxic insult. Furthermore, the ultimate molecular changes can occur following signal transduction initiated not only by damage, but also by specific receptor-mediated pathways. Finally, it is obvious that those ultimate molecular changes need to be defined in terms of biochemical markers and specific molecular mechanisms which would involve mediation both of degradative events and of events that are associated with inhibition of death and, therefore, with cell survival. Until that threshold is reached, it will remain difficult to tell whether cells die as a consequence of apoptosis, which implies possible therapeutic intervention, or as a consequence of necrosis, which implies that no intervention is possible after the initial damage. As the reader can see from the subsequent chapters of this book, investigators working in a wide range of fields within biology are struggling with the empirical problems of identification of specific molecular events that occur during apoptosis.

The concept of apoptosis may lead to challenges of conventional thought in toxicology, radiobiology, immunology, carcinogenesis, and developmental biology with respect to particular molecular mechanisms. To whatever extent these respective views may be altered, the consequence can be seen in terms of development of new means to modulate apoptosis, which thereby influence therapeutic efficacy of chemicals and radiation, cell-mediated immune function, or perhaps many physiological or pathological events that lead to cell death. This can be expected to take rational drug and therapy design to the next generation, as well as to introduce fundamentally new opportunities for the treatment and control of disease.

In the final analysis, this book has been prepared for the purpose of providing graduate students and scientists an introduction to the concept of apoptosis and an opportunity to consider how scientific hypothesis functions. These chapters have been organized to provide broad historical perspective and consideration of the current challenges regarding the molecular biology through essays and reviews written by some of the scientists who have been working with this new concept. Perhaps more importantly, there has been an attempt to consider the concept of apoptosis from the different points of view offered by researchers working in cancer therapeutics, immunology, neurobiology, invertebrate biology, and theoretical biology. These authors wish to provide their individual and sometimes divergent views within a single volume in order to help communicate the thought that apoptosis is a fundamentally important concept to all of biology and medicine. As such, apoptosis requires the greatest degree of critical analysis, for the object of this science is to understand the behavior of living cells.

Definition and Incidence of Apoptosis: An Historical Perspective

J.F.R. Kerr and B.V. Harmon

Department of Pathology, University of Queensland Medical School
Herston, Brisbane, Queensland 4006, Australia

The theme of this book is the hypothesis that apoptosis is a distinct type of cell death that differs fundamentally from degenerative death or necrosis in its nature and biological significance. Thus, there is evidence that apoptosis is an active process of gene-directed cellular self-destruction and that in most of the circumstances in which it occurs, it serves a biologically meaningful, homeostatic function. Necrosis, in contrast, is essentially accidental in its occurrence, being the outcome of severe injurious changes in the environment of the affected cells. Support for the apoptosis concept is provided by studies of the ultrastructure, biochemistry, and incidence of cell death. In this paper, we briefly survey these studies, adopting an historical perspective.

Morphological Definition of Apoptosis

Early evidence for the existence of two morphologically distinct types of cell death came from histochemical studies of changes in lysosomes in hepatic ischemia (Kerr 1965). Within hours of ligation of the portal vein branches supplying the left and median lobes of the rat liver, patches of confluent necrosis developed in these lobes around the terminal hepatic veins. The periportal parenchyma remained essentially viable, being sustained by the hepatic artery. However, over the succeeding weeks, it progressively shrank, the process continuing until a new balance was presumably established between parenchymal mass and residual blood supply; there was simultaneous hyperplasia of the other liver lobes. Throughout the period of parenchymal regression, scattered, individual hepatocytes were continually converted into small round cytoplasmic masses, some, but not all, of which contained minute specks of pycnotic chromatin. These structures were clearly a manifestation of cell death, but their histological appearance was different from that of the necrotic cells, and there was no associated inflammation, such as accompanied the necrosis. Histochemical staining for acid phosphatase indicated that lysosomes in the necrotic cells had ruptured and enzyme reaction product had been dispersed in the cytoplasm. However, lysosomes in the small round masses stained discretely, suggesting that they were still intact. Importantly, apparently identical rounded cytoplasmic masses were detected in very small numbers in the livers of healthy rats.

Electron microscopy was then used to study the development of the rounded masses (Kerr 1971). They were found to comprise membrane-enclosed cellular fragments containing crowded but structurally well-preserved organelles and, sometimes, pieces of compacted chromatin. They arose by a process of condensation and budding of hepatocytes. This morphological sequence (Fig. 1) has subsequently been recorded in a great variety of animal tissues (for illustrated reviews, see Wyllie et al. 1980; Kerr et al. 1987). It was originally referred to by the descriptive name "shrinkage necrosis" (Kerr 1971). However, use of the term necrosis for a phenomenon that occurs under physiological conditions seemed undesirable, and when evidence accumulated that cell death with this morphology plays an opposite role to mitosis in regulating tissue size, the name apoptosis was proposed to highlight its kinetic significance (Kerr et al. 1972). The word, like mitosis, is derived from the Greek and means "falling off," as of leaves from trees. An apparently similar process, at least as far as the nuclear changes are concerned, occurs in normal plants (Eleftheriou 1986). This suggests that apoptosis originated very early in biological evolution.

Cardinal Morphological Features of Apoptosis and Comparison with Necrosis

The earliest definitive changes in apoptosis that have been detected with the electron microscope are compaction of the

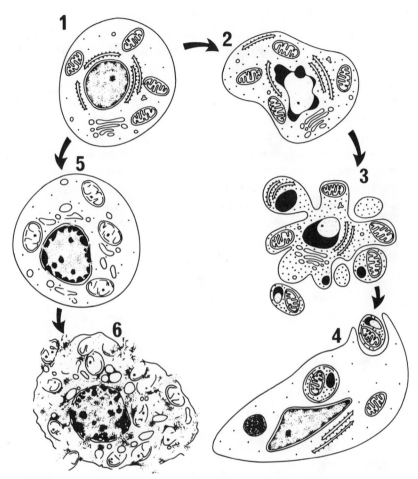

FIGURE 1 Diagram illustrating the sequential ultrastructural changes
in apoptosis (*right*) and necrosis (*left*). A normal cell is shown at 1.
The onset of apoptosis (2) is heralded by compaction and segregation
of chromatin into sharply delineated masses that lie against the nuclear envelope, condensation of the cytoplasm, and mild convolution
of the nuclear and cellular outlines. Rapid progression of the process
over the next few minutes (3) is associated with nuclear fragmentation and marked convolution of the cellular surface with the development of pedunculated protuberances. The latter then separate to produce membrane-bound apoptotic bodies, which are phagocytosed and
digested by adjacent cells (4). Signs of early necrosis in an irreversibly
injured cell (5) include clumping of chromatin into ill-defined masses,
gross swelling of organelles, and the appearance of flocculent densities in the matrix of mitochondria. At a later stage (6), membranes
break down and the cell disintegrates.

nuclear chromatin into sharply circumscribed, uniformly dense masses that abut on the nuclear envelope and condensation of the cytoplasm (Fig. 1). Continuation of condensation is accompanied by convolution of the nuclear and cellular outlines, and the nucleus often breaks up at this stage to produce discrete fragments (Figs. 1 and 2). The surface protuberances then separate with sealing of the plasma membrane, converting the cell into a number of membrane-bounded apoptotic bodies of varying size in which the closely packed organelles appear intact; some of these bodies lack a nuclear component, whereas others contain one or more nuclear fragments in which compacted chromatin is distributed either in peripheral crescents or throughout their cross-sectional area (Figs. 1 and 3). The extent of the nuclear fragmentation and cellular budding varies with cell morphology; they both tend to be relatively restricted in cells with a high nucleocytoplasmic ratio such

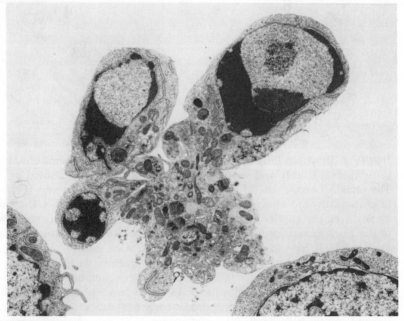

FIGURE 2 Apoptosis of murine NS-1 cell occurring spontaneously in culture. Note the discrete nuclear fragments with characteristic segregation of compacted chromatin, the crowding of organelles, and the marked convolution of the cellular surface.

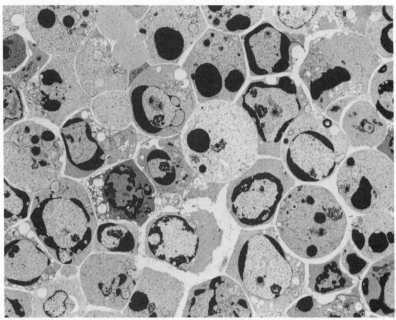

FIGURE 3 Apoptotic bodies derived from human BM 13674 lymphoma cells 4 hr after heating a log phase culture at 43°C for 30 min. Some nuclear fragments show peripheral crescents of compacted chromatin, whereas others are uniformly dense.

as thymocytes. In tissues, apoptotic bodies are rapidly taken up by adjacent cells and degraded within lysosomes (Figs. 1 and 4). A variety of cell types, including epithelial cells, may be involved in this tissue maintenance process. The surrounding cells close ranks, and the cell deletion is characteristically effected without disruption of overall tissue architecture. There is no inflammation. Some of the apoptotic bodies formed in single-layered epithelium may be extruded from its surface.

Phase-contrast microscopy of cells in culture shows that the condensation and budding to form apoptotic bodies are completed within several minutes (Russell et al. 1972; Sanderson 1976; Matter 1979). In vivo, the phagocytosis and digestion occur rapidly, and it is estimated that apoptotic bodies remain visible by light microscopy for only a few hours (Bursch et al. 1990). The process is thus often remarkably inconspicuous histologically.

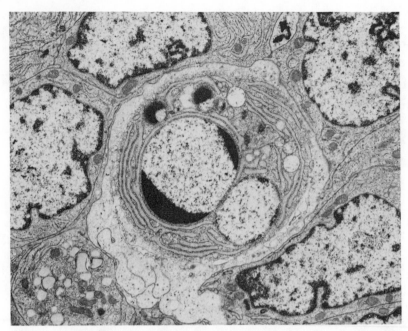

FIGURE 4 Apoptotic body containing well-preserved rough endoplasmic reticulum and four nuclear fragments, which has been phagocytosed by an intraepithelial macrophage in the rat ventral prostate 2
days after castration.

The ultrastructural appearances of necrosis are quite different (Trump et al. 1981), the main features being organelle
swelling and subsequent cellular disintegration (Figs. 1 and 5).
Although clumping of nuclear chromatin is usually evident,
the density of the masses is less uniform than in apoptosis,
and their edges are less sharply defined (cf. Figs. 5 and 2).
Early, potentially reversible swelling of an injured cell may be
due to direct membrane damage or to ATP depletion and disturbance of membrane pump activity. The point of no return
in irreversibly injured cells is marked by a gross increase in
membrane permeability, which is thought to be related in
many cases to activation of membrane-associated phospholipases by an increase in calcium (Farber et al. 1981; Trump et
al. 1984; Orrenius et al. 1989). Destruction of membrane integrity leads to release of lysosomal enzymes, which accelerate
disintegration (Trump et al. 1981). At a late stage, nuclear

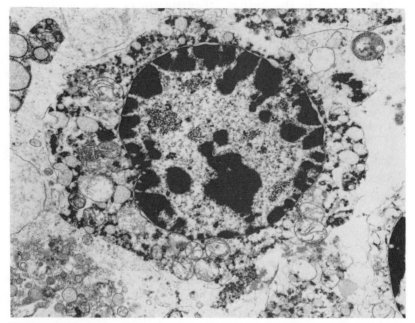

FIGURE 5 Spontaneous ischemic necrosis in the center of a murine P-815 tumor growing in muscle. Note the ill-defined edges of the clumps of compacted chromatin, the swelling of mitochondria, and the dissolution of membranes.

chromatin disappears, an event recognized by pathologists under the name karyolysis. Necrosis occurring in vivo is usually accompanied by exudative inflammation and, if large numbers of cells are involved, is often followed by the development of a scar.

Biochemical Mechanism of Apoptosis

The stereotyped morphology of apoptosis suggests a discrete phenomenon, and the rapid condensation of the cell with preservation of membranes favors its mechanism's being active rather than passive. Only a brief outline of the associated biochemical events is given here; they are covered in greater detail in later chapters.

Several features of apoptosis need to be borne in mind when biochemical studies are undertaken. First, the onset of the process in the individual members of a cell population is typically asynchronous (Wyllie et al. 1980). Second, whereas apoptotic bodies formed in cultures usually escape phagocytosis, thus obviating the difficulties posed by their rapid digestion in vivo, they spontaneously degenerate within several hours, a process sometimes referred to as secondary necrosis (Wyllie et al. 1980). It should be noted that they exclude vital dyes until this degeneration takes place (Sheridan et al. 1981).

The most extensively studied biochemical event in apoptosis involves the fate of nuclear DNA. Synchronously with the compaction of chromatin observed morphologically, double-strand cleavage occurs at the linker regions between nucleosomes to produce fragments that are multiples of approximately 185 bp (Wyllie 1980; Umansky 1982; Arends et al. 1990). These fragments can readily be demonstrated by agarose gel electrophoresis, wherein a characteristic "ladder" develops (Fig. 6). The method has been widely used to identify apoptosis. In necrosis, in contrast, the progressive disappearance of chromatin seen morphologically at a late stage of degeneration is accompanied by random DNA breakdown, a diffuse smear appearing in gels (Fig. 6).

The endonuclease presumed to be responsible for the internucleosomal DNA cleavage in apoptosis has not yet been purified and fully characterized (Arends et al. 1990). A candidate enzyme, which is present constitutively in the nuclei of rodent thymocytes, has been found to be calcium- and magnesium-dependent and to be inhibited by zinc (Cohen and Duke 1984). Cytosolic calcium levels increase during apoptosis of thymocytes (McConkey et al. 1988), and experiments showing that the process can be induced in these cells by calcium ionophores (Wyllie et al. 1984; Smith et al. 1989) and inhibited by calcium chelators (McConkey et al. 1988) suggest that the calcium elevation is involved in its initiation, possibly by directly activating the endonuclease (McConkey et al. 1990). The capacity for an elevation of intracellular calcium to induce thymocyte apoptosis depends, however, on additional factors. For example, phorbol esters in low concentration have been shown to switch the effect of calcium elevation from cell death to cell proliferation (Kizaki et al. 1989), probably through activation of protein kinase C (McConkey et al. 1990). Whether

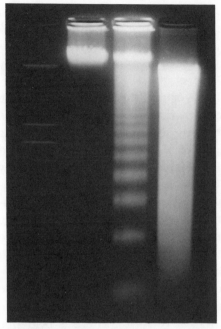

FIGURE 6 Agarose gel electrophoresis of DNA extracted from cultures of murine P-815 cells. Ethidium bromide stain; photographed under UV illumination. (Lane *1*) Molecular weight markers (DRIgest III); (lane *2*) log phase control culture; (lane *3*) culture showing extensive, morphologically typical apoptosis 8 hr after heating at 44°C for 30 min; (lane *4*) culture showing massive necrosis with karyolysis 72 hr after repeated freezing and thawing.

calcium plays a central role in initiating apoptosis in all types of cells is uncertain; recent evidence suggests that results obtained in experiments on thymocytes may not be universally applicable (Alnemri and Litwack 1990; Bansal et al. 1990). The fact that the putative apoptotic endonuclease is inhibited by zinc may explain both the reported suppression of apoptosis by this substance in cultures (Cohen and Duke 1984; Cohen et al. 1985) and the greatly increased rate of spontaneous apoptosis observed in the intestinal epithelium of zinc-deficient rats (Elmes and Gwyn Jones 1980).

A particularly interesting feature of apoptosis is its prevention in many situations by inhibitors of protein synthesis, such as cycloheximide (Lieberman et al. 1970; Ben-Ishay and

Farber 1975; Pratt and Greene 1976; Wyllie et al. 1984; Cohen et al. 1985; Yamada and Ohyama 1988). This has often been adduced as evidence for the active nature of apoptosis. Cyclo-heximide does not, however, suppress apoptosis in all circum-stances. It has no blocking effect on apoptosis induced by cytotoxic T lymphocytes (Duke et al. 1983), by mild hyper-thermia (Harmon et al. 1989), or by the fungal metabolite gliotoxin (Waring 1990). Moreover, cycloheximide itself induces apoptosis in rapidly proliferating cell populations such as normal intestinal crypts (Searle et al. 1975). It is possible that protein synthesis is essential for initiation of apoptosis by some types of stimuli, but that it is not an absolute require-ment for execution of the process. On the other hand, sub-stances such as cycloheximide may not completely abrogate protein synthesis (Martin et al. 1990), and experiments to determine whether certain proteins are synthesized when apoptosis occurs in their presence are needed.

One specific protein whose induction and activation have been clearly demonstrated to occur during apoptosis is tissue transglutaminase (Fesus et al. 1987, 1989). There is evidence that this enzyme brings about extensive cross-linking of pro-teins to produce a rigid shell under the bounding membranes of apoptotic bodies, thus helping to prevent leakage of their contents prior to phagocytosis by nearby cells (Fesus et al. 1989). Massive expression of the testosterone-repressed pros-tate message-2 gene (TRPM-2) has been observed to parallel enhanced apoptosis in several different tissues (Wadewitz and Lockshin 1988; Buttyan et al. 1989). However, the published micrographs suggest that the gene expression is not confined to the dying cells, and the precise relationship of the gene product to the occurrence of apoptosis remains to be determined (Bandyk et al. 1990).

The cellular condensation that characterizes apoptosis has been shown to be accompanied by an increase in density; this makes it possible to isolate apoptotic bodies by centrifugation on Percoll gradients (Wyllie and Morris 1982). The mechanism of the condensation does not, however, appear to have been systematically studied. Likewise, little is known about the cytoskeletal changes that must be involved in the cellular con-volution and budding (Wyllie et al. 1980).

The rapid phagocytosis of apoptotic bodies by nearby cells indicates an alteration in their plasma membranes. Recently, the vitronectin receptor, a heterodimer belonging to the cyto-adhesin family of integrins, has been implicated in the recognition of apoptotic neutrophils and lymphocytes by macrophages (Savill et al. 1990).

Incidence of Apoptosis

Knowledge of the incidence of apoptosis is based on studies using electron microscopy or DNA gel electrophoresis for its identification. In some cases, both of these methods have been used. The circumstances of its occurrence have been found to be basically different from those of necrosis (Wyllie 1981; Walker et al. 1988b). Thus, apoptosis accounts for deletion of cells in normal tissues and, where it occurs pathologically, an adaptive role for the cell death is usually evident. Necrosis, on the other hand, is invariably the result of a gross insult to the cell, such as severe hypoxic, ischemic, hyperthermic, or toxic damage; freezing and thawing; or the induction of membrane lesions by complement activation. There is only one area of potential overlap in the incidence of the two types of death, which is dealt with below under the heading of mild injury. We begin our survey of incidence with a consideration of the occurrence of apoptosis during normal embryonic development and metamorphosis, the one situation in which the importance of physiological cell death was explicitly recognized many years ago (Glücksmann 1951; Saunders 1966).

Cell death during normal development. Apoptosis plays a vital role in the embryonic development of higher vertebrates, being involved in such phenomena as deletion of interdigital webs (Hammar and Mottet 1971; Kerr et al. 1987), palatal fusion (Farbman 1968), and development of the intestinal mucosa (Harmon et al. 1984) and retina (Penfold and Provis 1986). The cell death responsible for regression of larval organs during metamorphosis of amphibia displays the classic morphological features of apoptosis (Kerr et al. 1974). However, developmental death occurring in larvae of the nematode *Caenorhabditis elegans*, although conforming to the overall ultrastructural pattern, is somewhat atypical in showing whorling of internal membranes and heightened autophagy (Robertson and Thomson 1982). The genetic regulation of cell death in this organism has been studied extensively (Yuan and Horvitz 1990). In insects and vertebrates, developmental death has been demonstrated to be under the influence of locally produced morphogens and circulating hormones (Hinchliffe 1981; Lockshin and Zakeri-Milovanovic 1984).

Cell death in normal adult tissues. Apoptosis occurs continuously in slowly proliferating cell populations such as hepatic (Kerr 1971; Benedetti et al. 1988) and adrenal cortical epithelium (Wyllie et al. 1973b) and in rapidly proliferating populations such as intestinal crypt epithelium (Potten 1977) and differentiating spermatogonia (Allan et al. 1987). In the former, apoptosis balances mitosis over a period of time (Wyllie et al. 1980), whereas in the latter, much of the mitosis is obviously related to loss of cells from the tissue by migration and shedding. There is evidence that the homeostatic regulation of normal tissue mass is effected by the cyclic production of growth factors and death factors, which induce mitosis and apoptosis, respectively (Lynch et al. 1986). Moreover, some growth factors have been shown to suppress apoptosis as well as to stimulate mitosis (Duke and Cohen 1986; Williams et al. 1990). It is not known, however, what determines the probability of an individual cell's dying at a particular time under steady-state conditions.

Apoptotic deletion of lymphocytes serves important normal functions in the immune system. Immature thymocytes have been shown to undergo apoptosis when stimulated via their T-cell receptor at a critical stage in their maturation, and it has been proposed that this is the mechanism responsible for elimination of autoreactive T-cell clones during the development of cellular immune self-tolerance (Shi et al. 1989; Smith et al. 1989). So-called tingible bodies, which are a prominent feature of germinal centers in lymphoid tissues reacting to antigenic stimulation, show the ultrastructure of apoptosis (Swartzendruber and Congdon 1963; Searle et al. 1982). Perhaps the cell death in this situation is involved in eliminating cells with inappropriate mutations generated during the process of somatic hypermutation responsible for fine-tuning of the affinity between antibody and antigen (Liu et al. 1989).

Apoptosis accounts for the cell deletion that accompanies a variety of normal involutional processes. These include ovarian follicular atresia (O'Shea et al. 1978); catagen involution of hair follicles (Weedon and Strutton 1981); and regression of the lactating breast after weaning (Walker et al. 1989), of the endometrium at estrus (Sandow et al. 1979), and of the adrenal cortex during the neonatal period (Wyllie et al. 1973a). In most of these examples of normal involution, the cell death is triggered by hormonal signals. However, in the case of catagen, the control mechanisms are still imperfectly understood (Hollis and Chapman 1987).

Senescent megakaryocytes that have released their platelets undergo apoptosis (Radley and Haller 1983), and aging neutrophil leukocytes are also eliminated in this way (Savill et al. 1989). Whether apoptosis is involved in the tissue changes associated with aging of the body as a whole remains to be determined (Goya 1986). An electron micrograph of late-passage fibroblasts by Brock and Hay (1971; Fig. 18 in their paper) suggests that in vitro cellular senescence may be accompanied by enhanced apoptosis. Finally in this context, it is of interest that terminal differentiation of optic lens fiber cells is associated with nuclear events similar to those occurring in apoptosis (Appleby and Modak 1977).

Cell death during pathological atrophy and regression of hyperplasia. As would be expected from what has been said about normal involution, apoptosis has been found to be involved in the pathological atrophy of endocrine-dependent tissues that follows artificial withdrawal of trophic hormonal stimulation, examples including atrophy of the prostate after castration (Kerr and Searle 1973; Kyprianou and Isaacs 1988; English et al. 1989) and of the adrenal cortex after suppression of ACTH secretion by glucocorticoid administration (Wyllie et al. 1973b). In the case of the thymus, administration of glucocorticoids causes atrophy directly by inducing apoptosis of thymocytes (Wyllie 1980). Apoptosis is, however, also involved in other types of atrophy, such as occur in the pancreas after ductular obstruction (Walker 1987) and in the renal parenchyma after ureteric ligation (Gobe and Axelsen 1987); in these latter situations, the proximate triggers of the cell death are unknown.

Where pathological enlargement or hyperplasia of a tissue has been produced by a mitogenic stimulus, reversion of the tissue to normal on withdrawal of the stimulus is effected by apoptosis. Well-documented examples include regression of hyperplasia of hepatocytes after cessation of administration of cyproterone acetate (Bursch et al. 1985) and of biliary ductules after relief of obstruction of the common bile duct (Bhathal and Gall 1985).

Altruistic cell suicide. In the instances of occurrence of apoptosis to be described in this section, it can be argued teleologically that the death serves a biologically useful role in eliminating cells whose survival might be harmful to the animal as a whole—for example, mutant or virus-infected cells.

Ionizing radiation induces apoptosis in proliferating cell populations such as the epithelium of intestinal crypts (Hugon and Borgers 1966; Potten 1977; Kerr and Searle 1980) and the spermatogonia of seminiferous tubules (Allan et al. 1987), and it also causes apoptosis of nonproliferating cells in lymphoid organs (Trowell 1966; Lucas and Peakman 1969; Zhivotovsky et al. 1981; Matyášová et al. 1984; Yamada and Ohyama 1988). Gross cellular depletion in these tissues after severe radiation exposure is likely, of course, to have adverse effects. However, under natural conditions, animals would rarely encounter large doses of radiation, and it is clearly desirable that the occasional cell with unrepaired radiation-induced DNA damage should be eliminated. In proliferating populations, persistence of such a cell might result in amplification of the genetic defect. This would have especially serious consequences in the case of spermatogonia, which give rise to gametes. The unique propensity for nonproliferating lymphocytes to undergo apoptosis after irradiation may be explained teleologically by the potential for mutant lymphoid cells to produce autoimmune disease (Cohen et al. 1985; Sellins and Cohen 1987). Phorbol esters, which are classic tumor-promoting agents, have been shown to inhibit radiation-induced apoptosis; at least part of their action in tumor promotion may thus depend on their allowing mutations to persist (Tomei et al. 1988). Certain chemical carcinogens induce apoptosis in their target tissues (Ronen and Heddle 1984; Walker et al. 1988a; Ijiri 1989), and many cancer-chemotherapeutic agents greatly enhance apoptosis in naturally rapidly proliferating populations of cells as well as in tumors (Philips and Sternberg 1975; Searle et al. 1975; Kuo and Hsu 1978; Ijiri and Potten 1987; see Tritton, this volume).

Cell death induced in vitro by T cells (Don et al. 1977; Sanderson and Glauert 1977; Matter 1979; Russell and Dobos 1980; Russell et al. 1982; Duke et al. 1983), K cells (Sanderson and Thomas 1977; Stacey et al. 1985), and NK cells (Bishop and Whiting 1983) has been shown to take the form of apoptosis, and apoptosis is observed in vivo in graft-versus-host disease (Kerr et al. 1987) and cellular immune rejection of allografts (Searle et al. 1977). A major function of the cellular immune system is the combating of viral infections. It is attractive to speculate that this system made opportunistic use during evolution of a preexisting cell-deletion mechanism that enabled it to eliminate virus-infected cells with minimal tissue disruption. The early fragmentation of DNA in apoptosis may be particularly important in viral containment, since it could halt viral replication and possibly also inactivate already assembled DNA virus within minutes of contact between a cytotoxic lymphocyte and an infected cell (Martz and Howell 1989). The general biological function of the DNA fragmentation may be the prevention of transfer of intact genetic material when apoptotic bodies are phagocytosed (Arends et al. 1990).

Cell death produced by mild injury. Here we consider the induction of apoptosis by agents that are also capable of producing necrosis, the type of cell death depending on the severity of the insult, not its qualitative nature.

In the case of in vitro hyperthermic cell damage, for example, it has been shown that there is a fairly abrupt change in the type of cell death induced when a critical heat load is exceeded. Thus, heating murine P-815 cells at 43°C or 44°C for 30 minutes causes rapid and exclusive enhancement of apoptosis, whereas heating the same cells at 46°C or 47°C causes massive necrosis (Harmon et al. 1990). After 45°C heating, both apoptosis and necrosis are increased. Certain toxins that produce zonal necrosis in the liver also enhance apoptosis in the nonnecrotic parenchyma (Kerr 1969; Reynolds et al. 1984). It is known that cells in the various zones of the liver lobules differ biochemically; those that are very susceptible to the respective toxins presumably all sustain severe damage and undergo necrosis, whereas damage to the more resistant cells is less severe, with some undergoing apoptosis.

Duvall and Wyllie (1986) have suggested that mild membrane damage caused by injurious agents might allow an influx of calcium sufficient to trigger apoptosis, but insufficient to extensively activate phospholipases with resultant necrosis. Even in the presence of injury, however, the induction of apoptosis may sometimes still be subject to homeostatic control. At the beginning of this chapter, we recounted how sustained, mild hepatic ischemia enhances apoptosis over a period of some weeks. Seventy years ago, Peyton Rous showed that the atrophy of the ischemic lobes in this experimental model is conditional on hyperplasia of the rest of the liver (Rous and Larimore 1920). Overriding homeostatic mechanisms may thus occasionally prevent apoptosis that would otherwise result from a mildly adverse cellular environment.

Spontaneous cell death in malignant tumors. Apoptosis occurs spontaneously in virtually all malignant tumors (Moore 1987), often markedly retarding their growth (Kerr and Searle 1972). Mild ischemia (Moore 1987), infiltration by cytotoxic lymphocytes (Curson and Weedon 1979), and release of tumor necrosis factor by macrophages (Sarraf and Bowen 1986) are probably involved in its causation in particular cases. However, much of the randomly distributed apoptosis in tumors may be a result of the operation in neoplastic cell populations of the autoregulatory mechanisms that control the size of normal tissues. It is of great interest that enhanced apoptosis is observed in preneoplastic foci during studies of chemical carcinogenesis in the liver as well as in the nodules and tumors that subsequently develop (Bursch et al. 1984; Columbano et al. 1984).

New methods for tipping the balance between apoptosis and mitosis in tumors have recently been reported (Szende et al.

1989; Trauth et al. 1989). The therapeutic implications of such findings are obvious.

CONCLUSIONS

Apoptosis was originally defined on the basis of its morphology. The hypothesis that it is a discrete phenomenon is supported by biochemical studies and its distinctive incidence. Elucidation of its biochemical mechanism has, however, only just begun. It is hoped that this book will stimulate further work in the field. Implicit in the proposal that apoptosis is an active process is the possibility of its *selective* regulation. The capacity to control apoptosis in diseased tissues would have major medical consequences.

ACKNOWLEDGMENTS

This work was supported by the Queensland Cancer Fund, the National Health and Medical Research Council of Australia, and the University of Queensland. We are grateful to Anne Corder and Clay Winterford for their assistance in producing the figures.

REFERENCES

Allan, D.J., B.V. Harmon, and J.F.R. Kerr. 1987. Cell death in spermatogenesis. In *Perspectives on mammalian cell death* (ed. C.S. Potten), p. 229. Oxford University Press, England.

Alnemri, E.S. and G. Litwack. 1990. Activation of internucleosomal DNA cleavage in human CEM lymphocytes by glucocorticoid and novobiocin. Evidence for a non-Ca^{2+}-requiring mechanism(s). *J. Biol. Chem.* **265:** 17323.

Appleby, D.W. and S.P. Modak. 1977. DNA degradation in terminally differentiating lens fiber cells from chick embryos. *Proc. Natl. Acad. Sci.* **74:** 5579.

Arends, M.J., R.G. Morris, and A.H. Wyllie. 1990. Apoptosis. The role of the endonuclease. *Am. J. Pathol.* **136:** 593.

Bandyk, M.G., I.S. Sawczuk, C.A. Olsson, A.E. Katz, and R. Buttyan. 1990. Characterization of the products of a gene expressed during androgen-programmed cell death and their potential use as a marker of urogenital injury. *J. Urol.* **143:** 407.

Bansal, N., A.G. Houle, and G. Melnykovych. 1990. Dexamethasone-induced killing of neoplastic cells of lymphoid derivation: Lack of early calcium involvement. *J. Cell. Physiol.* **143:** 105.

Benedetti, A., A.M. Jézéquel, and F. Orlandi. 1988. Preferential distribution of apoptotic bodies in acinar zone 3 of normal human and rat liver. *J. Hepatol.* **7:** 319.

Ben-Ishay, Z. and E. Farber. 1975. Protective effects of an inhibitor of protein synthesis, cycloheximide, on bone marrow damage induced by cytosine arabinoside or nitrogen mustard. *Lab. Invest.* **33:** 478.

Bhathal, P.S. and J.A.M. Gall. 1985. Deletion of hyperplastic biliary epithelial cells by apoptosis following removal of the proliferative stimulus. *Liver* **5:** 311.

Bishop, C.J. and V.A. Whiting. 1983. The role of natural killer cells in the intravascular death of intravenously injected murine tumour cells. *Br. J. Cancer* **48:** 441.

Brock, M.A. and R.J. Hay. 1971. Comparative ultrastructure of chick fibroblasts in vitro at early and late stages during their growth span. *J. Ultrastruct. Res.* **36:** 291.

Bursch, W., H.S. Taper, B. Lauer, and R. Schulte-Hermann. 1985. Quantitative histological and histochemical studies on the occurrence and stages of controlled cell death (apoptosis) during regression of rat liver hyperplasia. *Virchows Arch. B Cell Pathol.* **50:** 153.

Bursch, W., S. Paffe, B. Putz, G. Barthel, and R. Schulte-Hermann. 1990. Determination of the length of the histological stages of apoptosis in normal liver and in altered hepatic foci of rats. *Carcinogenesis* **11:** 847.

Bursch, W., B. Lauer, I. Timmermann-Trosiener, G. Barthel, J. Schuppler, and R. Schulte-Hermann. 1984. Controlled death (apoptosis) of normal and putative preneoplastic cells in rat liver following withdrawal of tumor promoters. *Carcinogenesis* **5:** 453.

Buttyan, R., C.A. Olsson, J. Pintar, C. Chang, M. Bandyk, P.-Y. Ng, and I.S. Sawczuk. 1989. Induction of the *TRPM-2* gene in cells undergoing programmed death. *Mol. Cell. Biol.* **9:** 3473.

Cohen, J.J. and R.C. Duke. 1984. Glucocorticoid activation of a calcium-dependent endonuclease in thymocyte nuclei leads to cell death. *J. Immunol.* **132:** 38.

Cohen, J.J., R.C. Duke, R. Chervenak, K.S. Sellins, and L.K. Olson. 1985. DNA fragmentation in targets of CTL: An example of programmed cell death in the immune system. *Adv. Exp. Med. Biol.* **184:** 493.

Columbano, A., G.M. Ledda-Columbano, P.M. Rao, S. Rajalakshmi, and D.S.R. Sarma. 1984. Occurrence of cell death (apoptosis) in preneoplastic and neoplastic liver cells. A sequential study. *Am. J. Pathol.* **116:** 441.

Curson, C. and D. Weedon. 1979. Spontaneous regression in basal cell carcinomas. *J. Cutaneous Pathol.* **6:** 432.

Don, M.M., G. Ablett, C.J. Bishop, P.G. Bundesen, K.J. Donald, J. Searle, and J.F.R. Kerr. 1977. Death of cells by apoptosis following attachment of specifically allergized lymphocytes in vitro. *Aust. J. Exp. Biol. Med. Sci.* **55:** 407.

Duke, R.C. and J.J. Cohen. 1986. IL-2 addiction: Withdrawal of growth factor activates a suicide program in dependent T cells. *Lymphokine Res.* **5:** 289.

Duke, R.C., R. Chervenak, and J.J. Cohen. 1983. Endogenous endonuclease-induced DNA fragmentation: An early event in cell-mediated cytolysis. *Proc. Natl. Acad. Sci.* **80:** 6361.

Duvall, E. and A.H. Wyllie. 1986. Death and the cell. *Immunol. Today* **7:** 115.

Eleftheriou, E.P. 1986. Ultrastructural studies on protophloem sieve elements in *Triticum aestivum* L. Nuclear degeneration. *J. Ultrastruct. Mol. Struct. Res.* **95:** 47.

Elmes, M.E. and J. Gwyn Jones. 1980. Ultrastructural studies of Paneth cell apoptosis in zinc deficient rats. *Cell Tissue Res.* **208:** 57.

English, H.F., N. Kyprianou, and J.T. Isaacs. 1989. Relationship between DNA fragmentation and apoptosis in the programmed cell death in the rat prostate following castration. *Prostate* **15:** 233.

Farber, J.L., K.R. Chien, and S. Mittnacht. 1981. The pathogenesis of irreversible cell injury in ischemia. *Am. J. Pathol.* **102:** 271.

Farbman, A.I. 1968. Electron microscope study of palate fusion in mouse embryos. *Dev. Biol.* **18:** 93.

Fesus, L., V. Thomazy, and A. Falus. 1987. Induction and activation of tissue transglutaminase during programmed cell death. *FEBS Lett.* **224:** 104.

Fesus, L., V. Thomazy, F. Autuori, M.P. Ceru, E. Tarcsa, and M. Piacentini. 1989. Apoptotic hepatocytes become insoluble in detergents and chaotropic agents as a result of transglutaminase action. *FEBS Lett.* **245:** 150.

Glücksmann, A. 1951. Cell deaths in normal vertebrate ontogeny. *Biol. Rev.* **26:** 59.

Gobe, G.C. and R.A. Axelsen. 1987. Genesis of renal tubular atrophy in experimental hydronephrosis in the rat. Role of apoptosis. *Lab. Invest.* **56:** 273.

Goya, R.G. 1986. Role of programmed cell death in the aging process: An unexplored possibility. *Gerontology* **32:** 37.

Hammar, S.P. and N.K. Mottet. 1971. Tetrazolium salt and electron-microscopic studies of cellular degeneration and necrosis in the interdigital areas of the developing chick limb. *J. Cell Sci.* **8:** 229.

Harmon, B., L. Bell, and L. Williams. 1984. An ultrastructural study on the "meconium corpuscles" in rat foetal intestinal epithelium with particular reference to apoptosis. *Anat. Embryol.* **169:** 119.

Harmon, B.V., J. Allen, G.C. Gobé, J.F.R. Kerr, R.J. Collins, and D.J. Allan. 1989. Hyperthermic cell killing in murine tumours with particular reference to apoptosis. In *Hyperthermic oncology 1988. Volume 1. Summary papers* (ed. T. Sugahara and M. Saito), p. 129. Taylor and Francis, London.

Harmon, B.V., A.M. Corder, R.J. Collins, G.C. Gobé, J. Allen, D.J. Allan, and J.F.R. Kerr. 1990. Cell death induced in a murine mastocytoma by 42-47°C heating in vitro: Evidence that the form of death changes from apoptosis to necrosis above a critical heat load. *Int. J. Radiat. Biol.* **58**: 845.

Hinchliffe, J.R. 1981. Cell death in embryogenesis. In *Cell death in biology and pathology* (ed. I.D. Bowen and R.A. Lockshin), p. 35. Chapman and Hall, London.

Hollis, D.E. and R.E. Chapman. 1987. Apoptosis in wool follicles during mouse epidermal growth factor (mEGF)-induced catagen regression. *J. Invest. Dermatol.* **88**: 455.

Hugon, J. and M. Borgers. 1966. Ultrastructural and cytochemical studies on karyolytic bodies in the epithelium of the duodenal crypts of whole body X-irradiated mice. *Lab. Invest.* **15**: 1528.

Ijiri, K. 1989. Apoptosis (cell death) induced in mouse bowel by 1,2-dimethylhydrazine, methylazoxymethanol acetate, and γ rays. *Cancer Res.* **49**: 6342.

Ijiri, K. and C.S. Potten. 1987. Cell death in cell hierarchies in adult mammalian tissues. In *Perspectives on mammalian cell death* (ed. C.S. Potten), p. 326. Oxford University Press, England.

Kerr, J.F.R. 1965. A histochemical study of hypertrophy and ischaemic injury of rat liver with special reference to changes in lysosomes. *J. Pathol. Bacteriol.* **90**: 419.

———. 1969. An electron-microscope study of liver cell necrosis due to heliotrine. *J. Pathol.* **97**: 557.

———. 1971. Shrinkage necrosis: A distinct mode of cellular death. *J. Pathol.* **105**: 13.

Kerr, J.F.R. and J. Searle. 1972. A suggested explanation for the paradoxically slow growth rate of basal-cell carcinomas that contain numerous mitotic figures. *J. Pathol.* **107**: 41.

———. 1973. Deletion of cells by apoptosis during castration-induced involution of the rat prostate. *Virchows Arch. B Cell. Pathol.* **13**: 87.

———. 1980. Apoptosis: Its nature and kinetic role. In *Radiation biology in cancer research* (ed. R.E. Meyn and H.R. Withers), p. 367. Raven Press, New York.

Kerr, J.F.R., B. Harmon, and J. Searle. 1974. An electron-microscope study of cell death in the anuran tadpole tail during spontaneous metamorphosis with special reference to apoptosis of striated muscle fibres. *J. Cell Sci.* **14**: 571.

Kerr, J.F.R., A.H. Wyllie, and A.R. Currie. 1972. Apoptosis: A basic biological phenomenon with wide-ranging implications in tissue kinetics. *Br. J. Cancer* **26**: 239.

Kerr, J.F.R., J. Searle, B.V. Harmon, and C.J. Bishop. 1987. Apoptosis. In *Perspectives on mammalian cell death* (ed. C.S. Potten), p. 93. Oxford University Press, England.

Kizaki, H., T. Tadakuma, C. Odaka, J. Muramatsu, and Y. Ishimura. 1989. Activation of a suicide process of thymocytes through DNA fragmentation by calcium ionophores and phorbol esters. *J. Immunol.* **143:** 1790.

Kuo, M.T. and T.C. Hsu. 1978. Bleomycin causes release of nucleosomes from chromatin and chromosomes. *Nature* **271:** 83.

Kyprianou, N. and J.T. Isaacs. 1988. Activation of programmed cell death in the rat ventral prostate after castration. *Endocrinology* **122:** 552.

Lieberman, M.W., R.S. Verbin, M. Landay, H. Liang, E. Farber, T.-N. Lee, and R. Starr. 1970. A probable role for protein synthesis in intestinal epithelial cell damage induced in vivo by cytosine arabinoside, nitrogen mustard, or X-irradiation. *Cancer Res.* **30:** 942.

Liu, Y.-J., D.E. Joshua, G.T. Williams, C.A. Smith, J. Gordon, and I.C.M. MacLennan. 1989. Mechanism of antigen-driven selection in germinal centres. *Nature* **342:** 929.

Lockshin, R.A. and Z. Zakeri-Milovanovic. 1984. Nucleic acids in cell death. In *Cell ageing and cell death* (ed. I. Davies and D.C. Sigee), p. 243. Cambridge University Press, England.

Lucas, D.R. and E.M. Peakman. 1969. Ultrastructural changes in lymphocytes in lymph-nodes, spleen and thymus after sublethal and supralethal doses of X-rays. *J. Pathol.* **99:** 163.

Lynch, M.P., S. Nawaz, and L.E. Gerschenson. 1986. Evidence for soluble factors regulating cell death and cell proliferation in primary cultures of rabbit endometrial cells grown on collagen. *Proc. Natl. Acad. Sci.* **83:** 4784.

Martin, S.J., S.V. Lennon, A.M. Bonham, and T.G. Cotter. 1990. Induction of apoptosis (programmed cell death) in human leukemic HL-60 cells by inhibition of RNA and protein synthesis. *J. Immunol.* **145:** 1859.

Martz, E. and D.M. Howell. 1989. CTL: Virus control cells first and cytolytic cells second? DNA fragmentation, apoptosis and the prelytic halt hypothesis. *Immunol. Today* **10:** 79.

Matter, A. 1979. Microcinematographic and electron microscopic analysis of target cell lysis induced by cytotoxic T lymphocytes. *Immunology* **36:** 179.

Matyášová, J., M. Skalka, and M. Čejková. 1984. On the degradation of chromatin to nucleosomes in the thymocytes of X-irradiated mice. *Folia Biologica* **30:** 123.

McConkey, D.J., S. Orrenius, and M. Jondal. 1990. Cellular signalling in programmed cell death (apoptosis). *Immunol. Today* **11:** 120.

McConkey, D.J., P. Hartzell, S.K. Duddy, H. Håkansson, and S. Orrenius. 1988. 2,3,7,8-Tetrachlorodibenzo-*p*-dioxin kills immature thymocytes by Ca^{2+}-mediated endonuclease activation. *Science*

242: 256.

Moore, J.V. 1987. Death of cells and necrosis of tumours. In *Perspectives on mammalian cell death* (ed. C.S. Potten), p. 295. Oxford University Press, England.

Orrenius, S., D.J. McConkey, G. Bellomo, and P. Nicotera. 1989. Role of Ca^{2+} in toxic cell killing. *Trends Pharmacol. Sci.* **10:** 281.

O'Shea, J.D., M.F. Hay, and D.G. Cran. 1978. Ultrastructural changes in the theca interna during follicular atresia in sheep. *J. Reprod. Fertil.* **54:** 183.

Penfold P.L. and J.M. Provis. 1986. Cell death in the development of the human retina: Phagocytosis of pyknotic and apoptotic bodies by retinal cells. *Graefe's Arch. Clin. Exp. Opthalmol.* **224:** 549.

Philips, F.S. and S.S. Sternberg. 1975. The lethal actions of antitumor agents in proliferating cell systems in vivo. *Am. J. Pathol.* **81:** 205.

Potten, C.S. 1977. Extreme sensitivity of some intestinal crypt cells to X and γ irradiation. *Nature* **269:** 518.

Pratt, R.M. and R.M. Greene. 1976. Inhibition of palatal epithelial cell death by altered protein synthesis. *Dev. Biol.* **54:** 135.

Radley, J.M. and C.J. Haller. 1983. Fate of senescent megakaryocytes in the bone marrow. *Br. J. Haematol.* **53:** 277.

Reynolds, E.S., M.F. Kanz, P. Chieco, and M.T. Moslen. 1984. 1,1-dichloroethylene: An apoptotic hepatotoxin? *Environ. Health Perspect.* **57:** 313.

Robertson, A.M.G. and J.N. Thomson. 1982. Morphology of programmed cell death in the ventral nerve cord of *Caenorhabditis elegans* larvae. *J. Embryol. Exp. Morphol.* **67:** 89.

Ronen, A. and J.A. Heddle. 1984. Site-specific induction of nuclear anomalies (apoptotic bodies and micronuclei) by carcinogens in mice. *Cancer Res.* **44:** 1536.

Rous, P. and L.D. Larimore. 1920. Relation of the portal blood to liver maintenance. A demonstration of liver atrophy conditional on compensation. *J. Exp. Med.* **31:** 609.

Russell, J.H. and C.B. Dobos. 1980. Mechanisms of immune lysis II. CTL-induced nuclear disintegration of the target begins within minutes of cell contact. *J. Immunol.* **125:** 1256.

Russell, J.H., V. Masakowski, T. Rucinsky, and G. Phillips. 1982. Mechanisms of immune lysis III. Characterization of the nature and kinetics of the cytotoxic T lymphocyte-induced nuclear lesion in the target. *J. Immunol.* **128:** 2087.

Russell, S.W., W. Rosenau, and J.C. Lee. 1972. Cytolysis induced by human lymphotoxin. Cinemicrographic and electron microscopic observations. *Am. J. Pathol.* **69:** 103.

Sanderson, C.J. 1976. The mechanism of T cell mediated cytotoxicity II. Morphological studies of cell death by time-lapse microcinematography. *Proc. R. Soc. Lond. B Biol. Sci.* **192:** 241.

Sanderson, C.J. and A.M. Glauert. 1977. The mechanism of T cell mediated cytotoxicity V. Morphological studies by electron micros-copy. *Proc. R. Soc. Lond. B Biol. Sci.* **198:** 315.

Sanderson, C.J. and J.A. Thomas. 1977. The mechanism of K cell (antibody-dependent) cell mediated cytotoxicity II. Characteristics of the effector cell and morphological changes in the target cell. *Proc. R. Soc. Lond. B Biol. Sci.* **197:** 417.

Sandow, B.A., N.B. West, R.L. Norman, and R.M. Brenner. 1979. Hor-monal control of apoptosis in hamster uterine luminal epithelium. *Am. J. Anat.* **156:** 15.

Sarraf, C.E. and I.D. Bowen. 1986. Kinetic studies on a murine sar-coma and an analysis of apoptosis. *Br. J. Cancer* **54:** 989.

Saunders, J.W. 1966. Death in embryonic systems. *Science* **154:** 604.

Savill, J., I. Dransfield, N. Hogg, and C. Haslett. 1990. Vitronectin receptor-mediated phagocytosis of cells undergoing apoptosis. *Na-ture* **343:** 170.

Savill, J.S., A.H. Wyllie, J.E. Henson, M.J. Walport, P.M. Henson, and C. Haslett. 1989. Macrophage phagocytosis of aging neutrophils in inflammation. Programmed cell death in the neutrophil leads to its recognition by macrophages. *J. Clin. Invest.* **83:** 865.

Searle, J., J.F.R. Kerr, and C.J. Bishop. 1982. Necrosis and apopto-sis: Distinct modes of cell death with fundamentally different sig-nificance. *Pathol. Annu.* **17(2):** 229

Searle, J., T.A. Lawson, P.J. Abbott, B. Harmon, and J.F.R. Kerr. 1975. An electron-microscope study of the mode of cell death in-duced by cancer-chemotherapeutic agents in populations of prolif-erating normal and neoplastic cells. *J. Pathol.* **116:** 129.

Searle, J., J.F.R. Kerr, C. Battersby, W.S. Egerton, G. Balderson, and W. Burnett. 1977. An electron microscopic study of the mode of donor cell death in unmodified rejection of pig liver allografts. *Aust. J. Exp. Biol. Med. Sci.* **55:** 401.

Sellins, K.S. and J.J. Cohen. 1987. Gene induction by γ-irradiation leads to DNA fragmentation in lymphocytes. *J. Immunol.* **139:** 3199.

Sheridan, J.W., C.J. Bishop, and R.J. Simmons. 1981. Biophysical and morphological correlates of kinetic change and death in a starved human melanoma cell line. *J. Cell Sci.* **49:** 119.

Shi, Y., B.M. Sahai, and D.R. Green. 1989. Cyclosporin A inhibits activation-induced cell death in T-cell hybridomas and thymo-cytes. *Nature* **339:** 625.

Smith, C.A., G.T. Williams, R. Kingston, E.J. Jenkinson, and J.J.T. Owen. 1989. Antibodies to CD3/T-cell receptor complex induce death by apoptosis in immature T cells in thymic cultures. *Nature* **337:** 181.

Stacey, N.H., C.J. Bishop, J.W. Halliday, W.J. Halliday, W.G.E. Cooksley, L.W. Powell, and J.F.R. Kerr. 1985. Apoptosis as the mode of cell death in antibody-dependent lymphocytotoxicity. *J. Cell Sci.* **74:** 169.

Swartzendruber, D.C. and C.C. Congdon. 1963. Electron microscope observations on tingible body macrophages in mouse spleen. *J. Cell Biol.* **19:** 641.

Szende, B., A. Zalatnai, and A.V. Schally. 1989. Programmed cell death (apoptosis) in pancreatic cancers of hamsters after treatment with analogs of both luteinizing hormone-releasing hormone and somatostatin. *Proc. Natl. Acad. Sci.* **86:** 1643.

Tomei, L.D., P. Kanter, and C.E. Wenner. 1988. Inhibition of radiation-induced apoptosis in vitro by tumor promoters. *Biochem. Biophys. Res. Comm.* **155:** 324.

Trauth, B.C., C. Klas, A.M.J. Peters, S. Matzku, P. Möller, W. Falk, K.-M. Debatin, and P.H. Krammer. 1989. Monoclonal antibody-mediated tumor regression by induction of apoptosis. *Science* **245:** 301.

Trowell, O.A. 1966. Ultrastructural changes in lymphocytes exposed to noxious agents in vitro. *Q. J. Exp. Physiol.* **51:** 207.

Trump, B.F., I.K. Berezesky, and A.R. Osornio-Vargas. 1981. Cell death and the disease process. The role of calcium. In *Cell death in biology and pathology* (ed. I.D. Bowen and R.A. Lockshin), p. 209. Chapman and Hall, London.

Trump, B.F., I.K. Berezesky, T. Sato, K.U. Laiho, P.C. Phelps, and N. De Claris. 1984. Cell calcium, cell injury and cell death. *Environ. Health Perspect.* **57:** 281.

Umansky, S.R. 1982. The genetic program of cell death. Hypothesis and some applications: Transformation, carcinogenesis, ageing. *J. Theor. Biol.* **97:** 591.

Wadewitz, A.G. and R.A. Lockshin. 1988. Programmed cell death: Dying cells synthesise a co-ordinated, unique set of proteins in two different episodes of cell death. *FEBS Lett.* **241:** 19.

Walker, N.I. 1987. Ultrastructure of the rat pancreas after experimental duct ligation I. The role of apoptosis and intraepithelial macrophages in acinar cell deletion. *Am. J. Pathol.* **126:** 439.

Walker, N.I., R.E. Bennett, and R.A. Axelsen. 1988a. Melanosis coli. A consequence of anthraquinone-induced apoptosis of colonic epithelial cells. *Am. J. Pathol.* **131:** 465.

Walker, N.I., R.E. Bennett, and J.F.R. Kerr. 1989. Cell death by apoptosis during involution of the lactating breast in mice and rats. *Am. J. Anat.* **185:** 19.

Walker, N.I., B.V. Harmon, G.C. Gobé, and J.F.R. Kerr. 1988b. Patterns of cell death. *Methods Achiev. Exp. Pathol.* **13:** 18.

Waring, P. 1990. DNA fragmentation induced in macrophages by gliotoxin does not require protein synthesis and is preceded by raised inositol triphosphate levels. *J. Biol. Chem.* **265:** 14476.

Weedon, D. and G. Strutton. 1981. Apoptosis as the mechanism of the involution of hair follicles in catagen transformation. *Acta Derm. Venereol.* **61:** 335.

Williams, G.T., C.A. Smith, E. Spooncer, T.M. Dexter, and D.R.

Taylor. 1990. Haemopoietic colony stimulating factors promote cell survival by suppressing apoptosis. *Nature* **343**: 76.

Wyllie, A.H. 1980. Glucocorticoid-induced thymocyte apoptosis is associated with endogenous endonuclease activation. *Nature* **284**: 555.

―――. 1981. Cell death: A new classification separating apoptosis from necrosis. In *Cell death in biology and pathology* (ed. I.D. Bowen and R.A. Lockshin), p. 9. Chapman and Hall, London.

Wyllie, A.H. and R.G. Morris. 1982. Hormone-induced cell death. Purification and properties of thymocytes undergoing apoptosis after glucocorticoid treatment. *Am. J. Pathol.* **109**: 78.

Wyllie, A.H., J.F.R. Kerr, and A.R. Currie. 1973a. Cell death in the normal neonatal rat adrenal cortex. *J. Pathol.* **111**: 255.

―――. 1980. Cell death: The significance of apoptosis. *Int. Rev. Cytol.* **68**: 251.

Wyllie, A.H., J.F.R. Kerr, I.A.M. Macaskill, and A.R. Currie. 1973b. Adrenocortical cell deletion: The role of ACTH. *J. Pathol.* **111**: 85.

Wyllie, A.H., R.G. Morris, A.L. Smith, and D. Dunlop. 1984. Chromatin cleavage in apoptosis: Association with condensed chromatin morphology and dependence on macromolecular synthesis. *J. Pathol.* **142**: 67.

Yamada, T. and H. Ohyama. 1988. Radiation-induced interphase death of rat thymocytes is internally programmed (apoptosis). *Int. J. Radiat. Biol.* **53**: 65.

Yuan, J. and H.R. Horvitz. 1990. The *Caenorhabditis elegans* genes *ced-3* and *ced-4* act cell autonomously to cause programmed cell death. *Dev. Biol.* **138**: 33.

Zhivotovsky, B.D., N.B. Zvonareva, and K.P. Hanson. 1981. Characteristics of rat thymus chromatin degradation products after whole-body X-irradiation. *Int. J. Radiat. Biol.* **39**: 437.

The Significance of Cell Death

J. Michaelson

*Department of Pathology, Harvard Medical School
and Massachusetts General Hospital Cancer Center
Charlestown, Massachusetts 02129*

Why is there cell death? Why, especially during development, is there such a luxuriant waste of cells? Why doesn't the embryo "do things right in the first place?" The very fact of cell death raises fundamental questions about the way in which developmental organization is created and maintained. Does development, in fact, proceed in a fashion that we would consider to be logical and efficient, or might the generation of the many different types of cells that appear early in life be a fundamentally chaotic process, which requires cell death to create order?

Till has pointed out that there are two types of cellular events which occur during development, to which he has given the names "Lamarckian" and "Darwinian" (Till 1981). The term Lamarckian is used to describe those occasions in which cells respond to signals in their environments and modify the expression of their genomes, whereas Darwinian describes those processes in which cells, having already become committed to particular patterns of gene expression, are selected to live or die. Both Darwinian and Lamarckian processes can generate the sort of developmental order that is seen to arise during embryonic life. The research that my colleagues and I have carried out in the last few years has led us to the belief that Darwinian processes provide the dominant organizing force in development (Michaelson 1987, 1989, and in prep.). As such, we have come to the view that cell death is more a cause, than a result, of developmental organization.

In our laboratory, we have been interested in the synthesis of the proteins of the blood, the so-called plasma proteins. These are produced either in the lymphoid organs of the immune system, which produce the immunoglobulins, or in the

Apoptosis: The Molecular Basis of Cell Death

parenchyma of the liver, which makes most of the rest. The synthesis of the immunoglobulins has a number of features, which, as I describe below, are used not only by the immune system, but may also be at work in regulating the expression of the plasma proteins produced by the liver. Two features of antibody production are central to the control of the immune response. *First*, antibody synthesis is the prototype of a Darwinian process, that is, a developmental process organized by cellular selection. The presence of a foreign substance (an "antigen") in the body causes the division of those few lymphocytes, which happen, by chance, to express a complementary immunoglobulin. Antibody-producing cells are selected by the immune system, and this process is known to immunologists as clonal selection (Jerne 1955; Burnet 1959; Lederberg 1959). *Second*, immunoglobulin gene expression is a stochastic, or random, process. Whether an individual lymphocyte expresses this or that V-region gene is essentially a matter of chance.

The random or stochastic property of immunoglobulin gene expression is now an accepted feature of the immune response. Historically, this has been detected by a number of methods, but one of the first indications of this randomness was the identification of the phenomenon of allelic exclusion (Pernis et al. 1965; Cebra et al. 1966). Being diploid organisms, vertebrate animals contain two homologous copies of each autosomal gene. However, for immunoglobulin genes, the two homologous copies are never expressed simultaneously in the same cell. Technically, allelic exclusion is detected with antisera that identify genetic variants, or allotypes, of immunoglobulin genes. Allelic exclusion can be seen in animals heterozygous for an immunoglobulin allotypic polymorphism. By using antiallotypic sera, individual lymphocytes in these animals are found to express one or the other allotype, but never both. Allelic exclusion follows from the stochastic feature of immunoglobulin gene expression, by the following reasoning: If the expression of each immunoglobulin gene is random, there is no reason to expect the two homologous alleles of a gene to be activated coordinately. If we hold two dice, one in each hand, and throw them both, there is no reason to expect both to come up the same.

The stochastic property of immunoglobulin genes may seem

to be strange and wasteful. However, this stochastic feature is essential to the proper functioning of the immune system, because it generates the diversity of lymphocytes upon which clonal selection can act.

The liver produces most of the hundred or so plasma proteins not made by the immune system. This tissue is composed predominantly of one type of cell, known as the hepatocyte or parenchymal cell, which, in conventional histological preparations, appears to be homogeneous. We have found, however, that functionally these cells are highly heterogeneous and specialized. As we show, this heterogeneity appears to be central to the regulation of plasma protein synthesis by the liver. We detected the heterogeneity of hepatocytes by carrying out immunofluorescence analyses on frozen sections of mouse and rat liver (Fig. 1) (Michaelson 1989 and in prep.). Albumin is seen in about 1% of hepatocytes, whereas transferrin, fibrinogen, α_2-macroglobulin, complement-component-C3, and α-fetoprotein are each present in separate hepatocytes, although in somewhat lower numbers. By carrying out two-color immunofluorescence experiments, we found that each of these populations of cells is separate. Plasma protein-containing cells are present in the liver both individually and in clusters, which are probably clonal. These cells are scattered throughout the liver without any obvious regional orientation.

The central point of these results is that, like lymphocytes in the immune systems, hepatocytes are functionally specialized, in which each protein appears to be produced by a separate population of cells. This heterogeneity appears to be used by the liver in much the same way that it is used by the immune system.

We carried out immunofluorescence reactions with antisera produced in rabbits, sheep, and goats, as well as with a mouse antiserum that reacts with one genetic variant of mouse albumin. There are two allelic forms of albumin in the mouse, named Alb^a and Alb^c (Petras 1972). We immunized Alb^a mice with Alb^c albumin, and produced an antiserum, which reacts with albumin of Alb^c genotype. This anti-Alb^c serum reacts with albumin-containing cells (recognized by rabbit anti-mouse albumin) when tested against the livers of Alb^c mice. It does not react at all against cells in the livers of Alb^a mice.

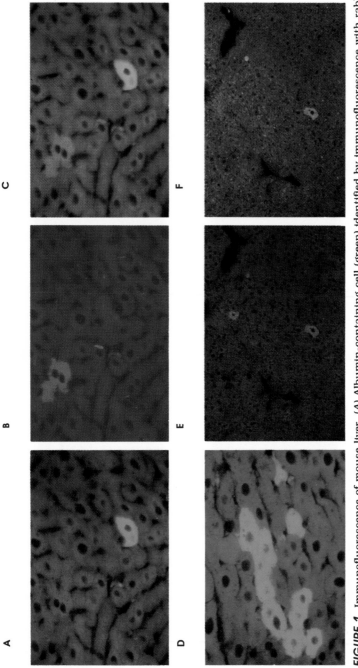

FIGURE 1 Immunofluorescence of mouse liver. (A) Albumin-containing cell (green) identified by immunofluorescence with rabbit anti-mouse albumin. (B) C3-containing cell in same field (red) identified by goat anti-mouse C3. (C) Double exposure of same fields as A and B, showing simultaneously albumin (green) and C3 (red) containing cells. A large series of similar experiments established the separate identity of cells containing albumin, transferrin, C3, fibrinogen, and α_2-macroglobulin. (D) Cluster of albumin-containing cells. (E,F) Immunofluorescence of liver from an Alb^a/Alb^c heterozygous mouse. The reactivity of rabbit anti-mouse albumin is revealed in red; the reactivity of mouse anti-Alb^c is seen in green.

When we examined the livers of Alb^a/Alb^c heterozygous mice, we found a very remarkable thing: The albumin locus appears to display allelic exclusion, exactly like that known for immunoglobulin genes. This could be seen in the observation that only about half of the albumin-containing cells (identified by rabbit anti-mouse albumin antiserum) express the Alb^c allele (identifiable by anti-Alb^c) in the livers of Alb^a/Alb^c mice (Fig. 1). The most straightforward, but perhaps most surprising, conclusion from this observation is that the expression of the albumin gene is a stochastic process.

As in the case of immunoglobulin genes, the stochastic feature of albumin gene expression strikes us as wasteful and illogical. However, as in the case of immunoglobulin gene expression, the usefulness of this random property can be seen in its capacity to generate the heterogeneity of hepatocytes present in the liver.

As mentioned previously, plasma protein-containing cells are present in the liver both as individual cells and in groups of adjacent cells, which we suspected might be clonal. The analysis of these groups of albumin-containing cells in Alb^a/Alb^c heterozygous mice substantiated the idea that these clusters are clones. We reached this conclusion from the observation that although some clusters in these mice are Alb^{c+} and some are Alb^{c-}, *within* each cluster there is no variation. We examined over 80 clusters containing approximately 400 cells from four separate heterozygous mice and did not find even one cluster that contained both Alb^{c+} and Alb^{c-} cells. Presumably, each cluster is a clonal group, derived from a single cell committed either to express the Alb^c gene or not, and this property is inherited by the progeny of that cell.

To summarize, these observations suggest two conclusions. *First*, the activation of the albumin gene appears to be a stochastic process. *Second*, the state of gene expression appears to be cell-heritable; once a hepatocyte has become committed to express an albumin gene, this property is passed along to the clonal progeny of that cell. The cell-heritable feature of gene expression is, of course, not unique to the albumin gene but is a widespread observation in biology, and there have been a number of suggestions as to the mechanism of this process. Among the hypotheses that have been put forward to explain cell-heritable changes in gene expression are methylation, supercoiling, protein-DNA interactions that are perpetuated through replication, and self-propagating hypersensitivity sites (Riggs 1989). As to the role of the stochastic activation step in the final determination of phenotype, I suspect that the stochastic step is only one of several components of gene expression. It seems likely to me that after gene activation occurs, there are additional levels of control, mediated by regulatory proteins interacting with cis-acting sequences, as have been suggested by many studies in the last few years.

The apparent stochastic feature of albumin gene activation, surprising as it may be, has precedents in other systems. Allelic exclusion has been observed, not only for immunoglobulin genes, but also for genes coding for phenotypes as diverse as skin coat color and erythrocyte hemoglobin. For coat color genes, this allelic exclusion is seen as a patchy or variegated appearance of the pelt of heterozygous animals. Although in several of these cases the coat color genes are on the X chromosome and the variegation can be ascribed to random X-chromosome inactivation, most examples of coat color variegation are autosomal. This variegation has been observed for coat color genes on half a dozen autosomes in the mouse, as well as for a variety of autosomal genes in guinea pigs, cattle, rabbits, dogs, and pigs (for references, see Michaelson 1987). For globin genes, allelic exclusion has been observed in erythroid precursors from children heterozygous for one of the fetal globin genes. Kidoguchi and his colleagues made this observation when they found that each of the two allelic forms of a fetal globin gene are not expressed coordinately in the progeny of single cells (Kidoguchi et al. 1980).

Stochastic gene expression has also been observed in a number of cases where the process of gene activation has been followed in single cells. This has been observed for the generation of hematopoietic stem cells by Till and McCollough (1980), for melanogenesis by Bennett (1983), for globin gene expression by Housman, Levinson, Gusella, Orkin, and Leder (Orkin et al. 1975; Gusella et al. 1976; Levinson and Housman 1981), for myogenesis by Nadal-Ginard (1978), and for terminal differentiation by Smith and Whitney (1980).

I turn now from the apparent functional heterogeneity of

hepatocytes, and the cause of this heterogeneity, to the manner in which this heterogeneity appears to be utilized by the liver to control the overall output of plasma proteins. We have collected a great deal of data on the quantity of cells of each

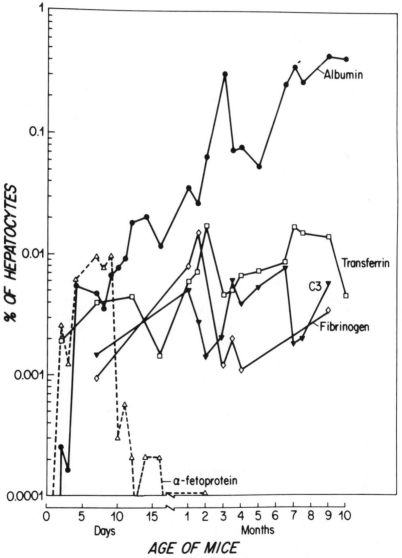

FIGURE 2 Percentage of cells containing plasma proteins in the livers of mice of various ages. (Data from J. Michaelson, in prep.).

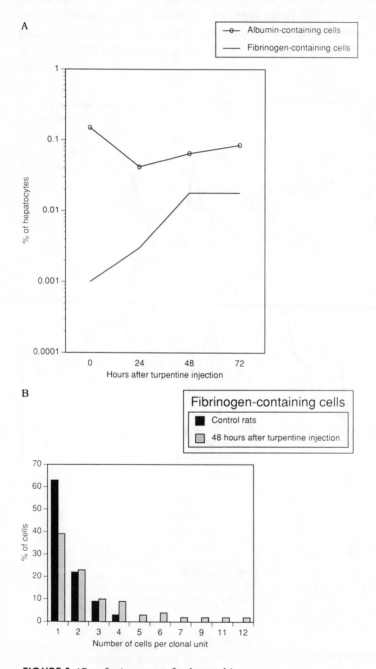

FIGURE 3 (*See facing page for legend.*)

type (Figs. 2 and 3) (Michaelson 1989 and in prep.). These data suggest that the functional status of the liver is determined by the cellular mix among the specialized types of hepatocytes and that changes in function may be mediated by changes in this mix by selective cell growth and death. For example, we have found that in the adult mouse liver, albumin-containing cells constitute somewhat less than 1% of hepatocytes. About one-tenth of this number of cells produce transferrin, and one-thirtieth of this number produce C3 (Fig. 2). These values roughly reflect the relative rates of synthesis of these three proteins.

The liver appears to be able to change its plasma protein output by changing the mix of specialized cells by cell growth and death. For example, early in life, α-fetoprotein is the principal plasma protein, but soon after birth, there is a gradual switch from α-fetoprotein synthesis to albumin synthesis (Abelev 1971; Nahon 1987). We have found that there is an equivalent change in the number of cells containing each protein (Fig. 2). There are considerable numbers of α-fetoprotein-containing cells in the livers of neonatal rats and mice; the number of these cells declines as the animals mature, while the number of albumin-containing cells rises from being very rare to being the most prevalent plasma protein-containing cell in the liver. This is apparently due to the fact that α-fetoprotein-containing cells divide most rapidly in the livers of newborn animals, as could be seen by the average number of cells contained in clonal clusters of α-fetoprotein-containing cells. Apparently, these cells divide more slowly as mice mature, as shown by a smaller number of cells in those clonal groups. The converse is true for albumin-containing cells. Thus, the switchover from a predominance of α-fetoprotein-containing cells to a predominance of albumin-containing cells appears to be due to differential cell growth.

FIGURE 3 (A) Change in the number of albumin- and fibrinogen-containing cells after an acute-phase response induced in rats by an injection of turpentine. (B) Distribution of fibrinogen-containing hepatocytes into single cells and clusters of various sizes in control rats and in rats 48 hr after induction of the acute-phase response. These data illustrate the increase in the average cluster size of fibrinogen-containing cells, apparently by cell division.

During adult life, there are also changes in plasma protein synthesis, particularly during inflammation. For example, Schreiber has shown that the inflammation induced in rats by an injection of turpentine causes, over a 2-day period, a slight decrease in albumin synthesis and a five- to tenfold increase in fibrinogen synthesis (Schreiber et al. 1989). We have found an equivalent slight reduction in the number of albumin-containing cells and a sixfold increase in the number of fibrinogen-containing cells (Fig. 3A). Furthermore, it appears that the sixfold increase in the number of fibrinogen-containing cells is due to cell division, as shown by the increase in the average size of the clusters of fibrinogen-containing cells (Fig. 3B). We have recently found direct evidence that the acute-phase response is accompanied by considerable cell growth in the liver, by visualizing dividing cells that take up the DNA precursor bromodeoxyuridine.

What can we conclude from the data collected so far? Our picture is far from complete, but we can make some generalizations, if only to frame testable hypotheses. The hepatocytes that comprise the liver appear to be heterogeneous in that each plasma protein is seen in a separate set of hepatocytes. This heterogeneity appears to be generated by stochastic gene activation. Change in the overall synthesis of each plasma protein may be mediated by selective growth and death among these specialized cells in the liver. In short, plasma protein synthesis appears to fit our description of a Darwinian process; that is, a process organized by cellular selection among a diverse collection of specialized cells.

What are the general lessons to be drawn from these observations? Does the process of plasma protein synthesis exemplify a general phenomenon in biology? The signs of cell growth and death are seen everywhere in embryology (Kerr et al. 1972, 1987; Beaulaton and Lockshin 1982). Glucksmann enumerated 74 separate examples of embryonic cell death alone (Glucksmann 1950). The cell death that accompanies development is not incidental but represents a highly refined cellular mechanism (Kerr et al. 1980; Tomei et al. 1988). Might the differential cell growth and death seen in development not just be the result of morphogenesis, but the *cause* of morphogenesis? Selective growth and death of cells is such a common feature of embryogenesis that it is easy to overlook the capacity of such processes to mold the outcome of development itself. For example, the liver starts out as just a few cells in the hepatic diverticulum of a 7-day-old mouse embryo; these cells grow much more rapidly than the rest of the embryo, generating, by this selective cell growth, an organ which by day 10 is recognizable as the liver by size and shape (Theiler 1989). In short, the liver comes into being by the selective growth of its precursors. Such events occur repeatedly in development (Michaelson 1987).

Within organs themselves, the process of the generation of form, or morphogenesis, can often be ascribed to cell selection. One of the best illustrations of such a process is the development of one system, the vertebrate limb (Fig. 4) (Michaelson 1987). Limb development begins in the embryo as a small outgrowth called the limb bud, composed of a boundary of ectoderm enclosing mesodermal cells (Hinchliffe and Johnson 1980). After a period of growth, condensations of bone-forming and muscle-forming cells appear roughly in the general pattern of bones and muscles that will constitute the mature limb. It appears unlikely that limb morphogenesis occurs by a Lamarckian process, in which the undifferentiated mesoder-

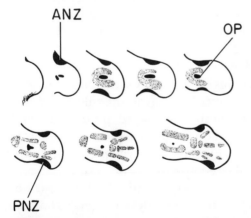

FIGURE 4 Development of the chick wing, showing areas of highest cell death (black) and areas of chondrogenic condensation (gray). (ANZ) Anterior necrotic zone; (PNZ) posterior necrotic zone; (OP) opaque patch. (Reprinted, with permission, from Michaelson 1987).

mal cells are instructed to become muscle or bone (by diffusible "morphogens" or other signals), for the simple reason that undifferentiated mesenchymal cells with potential to become either myogenic or chondrogenic do not exist in the limb. In fact, the two cell types appear to be derived from entirely separate, noninterconvertible cell lineages that migrate into the limb from different regions of the embryo (Abbott et al. 1974; Christ et al. 1977; for additional references, see Michaelson 1987). It seems that, like the parenchyma of the liver, the mesoderm of the limb bud is composed of a salt-and-pepper mixture of cells with separate potentialities.

What then, accounts for the appearance of limb structure? As in the case of the liver, we can account for limb morphogenesis by a Darwinian process of cellular selection. We see the signs of this selective process in the massive amounts of cell death throughout the developing limb. Large areas of this cell death are seen at the front and back of the limb bud (the anterior and posterior necrotic zones) and in the central area (the opaque patch) (Fig. 4) (Hinchliffe and Johnson 1980). We suspect that this cell death is the selective elimination of chondrogenic cells, since bone will later appear only in those areas where cell death is not seen. It might be said, then, that cell death does not arise as a result of limb development; limb development arises as a result of cell death (Michaelson 1987).

The immune system illustrates how a biochemical response can be finely tuned by a selective process, but is such a process capable of shaping development in three dimensions, as is seen in the limb? Fortunately, there have been several theoretical studies that suggest that such an outcome is, in fact, very much a possibility. In one of the nicest of such studies, Swindale asked whether a striped pattern, such as seen in a zebra's skin or in the visual cortex, can be generated by cellular selection alone (Fig. 5). He developed a computer-simulated model of cellular competition and, starting from a random mixture of cells interacting through diffusible substances, found that a highly patterned developmental outcome may result (Swindale 1980).

The features of morphogenesis driven by cell selection, and particularly cell death, are not unique to the limb, the liver (Bursch et al. 1985; Columbano et al. 1985), or the immune

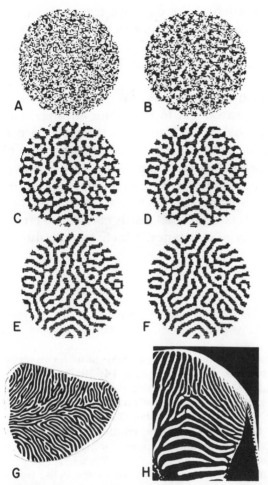

FIGURE 5 Computer simulation of the spontaneous emergence of striped form in a field of cells or synapses arising through a process of selection (*A–F*). Comparable examples from the natural world: (*G*) ocular dominance stripes in the visual cortex; (*H*) stripes on the rear flank of a zebra. (Redrawn from Swindale 1980.)

system (Duke 1989; Trauth et al. 1989), but appear to be a widespread feature in development. Common examples of this include regression of the Wolfian or Müllerian ducts, destruction of larval structures during tissue reformation in insect and amphibian metamorphosis (Wadewitz and Lockshin 1988), closure of the palate, invagination of the optic cup, shaping of the nose, creation of the digestive tract and heart, formation of the nervous system (Oppenheim and Nunez 1982; Ashwell and Watson 1983; Johnson et al. 1989), and remodeling of the uterine epithelium (Rotello et al. 1989) and prostate (Buttyan et al. 1989).

I began this chapter by posing the question, "Why is there cell death?" Perhaps we can begin to formulate an answer to this question. We can look to the role that cell selection can play in the generation of developmental order. The process of selective growth and death has great simplicity, but it is also capable of great versatility and precision. If, as we suspect, development is a Darwinian process of cellular selection, then the meaning of cell death is clear: Cell death may act to create order out of the diversity of cell types, which emerge during embryonic life.

REFERENCES

Abbott, J., J. Schiltz, S. Dienstman, and H. Holtzer. 1974. The phenotypic complexity of myogenic clones. *Proc. Natl. Acad. Sci.* **71:** 1506.

Abelev, G.I. 1971. Alpha-fetoprotein in oncogenesis and its association with malignant tumors. *Adv. Cancer Res.* **14:** 2915.

Ashwell, K.W. and C.R.R. Watson. 1983. The development of facial motoneurons in the mouse-neuronal death and the innervation of the facial muscles. *J. Embryol. Exp. Morphol.* **77:** 117.

Beaulaton, J. and R.A. Lockshin. 1982. The relation of programmed cell death to development and reproduction: Comparative studies and an attempt at classification. *Int. Rev. Cytol.* **79:** 215.

Bennett, D.C. 1983. Differentiation in mouse melanoma cells: Initial reversibility and an on-off stochastic model. *Cell* **34:** 445.

Burnet, F.M. 1959. *The clonal selection theory of acquired immunity.* Vanderbilt University Press, Nashville, Tennessee.

Bursch, H., S. Taper, B. Lauer, and R. Schulte Hermann. 1985. Quantitative histological and histochemical studies on the occurrence and stages of controlled cell death (apoptosis) during regression of rat liver hyperplasia. *Virchows Arch. B. Cell Pathol.* **50:** 153.

Buttyan, R., C.A. Olsson, J. Pintar, C. Chang, M. Bandyk, P.Y. Ng, and I.S. Sawczuk. 1989. Induction of the TRPM-2 gene in cells undergoing programmed death. *Mol. Cell. Biol.* **9:** 3473.

Cebra, J.J., J.E. Colberg, and S. Dray. 1966. Rabbit lymphoid cells differentiated with respect to heavy polypeptide chains and to allotypic markers Aa1 and Aa2. *J. Exp. Med.* **123:** 547.

Christ, B., H.J. Jacob, and M. Jacob. 1977. Experimental analysis of the origin of the wing musculature in avian embryos. *Anat. Embryol.* **150:** 171.

Columbano, A., G.M. Ledda-Columbano, P.P. Coni, G. Faa, C. Liguori, G. Santa Cruz, and P. Pani. 1985. Occurrence of cell death (apoptosis) during the involution of liver hyperplasia. *Lab. Invest.* **52:** 670.

Duke, R.C. 1989. Self recognition by T cells. I. Bystander killing of target cells bearing syngeneic MHC antigens. *J. Exp. Med.* **170:** 59.

Glucksmann, A. 1950. Cell death in normal vertebrate ontogeny. *Biol. Rev.* **26:** 59.

Gusella, J., R. Geller, B. Clarke, V. Weeks, and D. Housman. 1976. Commitment to erythroid differentiation by Friend erythroleukemia cells: A stochastic analysis. *Cell* **9:** 221.

Hinchliffe, J.R. and D.R. Johnson. 1980. *The development of the vertebrate limb.* Clarendon Press, Oxford, England.

Jerne, N.K. 1955. The natural selection theory of antibody formation. *Proc. Natl. Acad. Sci.* **41:** 849.

Johnson, E.M., Jr., J.Y. Chang, T. Koike, and D.P. Martin. 1989. Why do neurons die when deprived of trophic factor? *Neurobiol. Aging* **5:** 549.

Kerr, J.F.R., A.H. Wyllie, and A.R. Currie. 1972. Apoptosis: A basic biological phenomenon with wide-ranging implications in tissue kinetics. *Br. J. Cancer* **26:** 239.

Kerr, J.F.R., J. Searle, B.V. Harmon, and C.J. Bishop. 1987. Apoptosis. In *Perspectives on mammalian cell death* (ed. C.S. Potten), p. 93. Oxford University Press, Oxford.

Kidoguchi, K., M. Ogawa, J.D. Karam, R.G. Schneider, and U. Carpentieri. 1980. Biosynthesis of hemoglobin F. malta-I in culture by adult circulating erythropoietic precursors. *Blood* **55:** 334.

Lederberg, J. 1959. Genes and antibodies. *Science* **129:** 1649.

Levenson, R. and D. Housman. 1981. Commitment: How do cells make the decision to differentiate? *Cell* **25:** 5.

Michaelson, J. 1987. Cell selection in development. *Biol. Rev.* **62:** 115.

―――. 1989. Heterogeneity of liver parenchymal cells. *Cell Differ.* (suppl.). **27:** S1.

Nadal-Ginard, B. 1978. Commitment, fusion and biochemical differentiation of a myogenic cell line in the absence of DNA synthesis. *Cell* **15:** 855.

Nahon, J.L. 1987. The regulation of albumin and α-fetoprotein gene expression in mammals. *Biochimie* **69:** 445.

Oppenheim, R.W. and R. Nunez. 1982. Electrical stimulation of hind limb bud increases neuronal cell death in chick embryo. *Nature* **259:** 57.

Orkin, S.H., F.I. Harosi, and P. Leder. 1975. Differentiation in erythroleukemia cells and their somatic hybrids. *Proc. Natl. Acad. Sci.* **72:** 98.

Pernis, B., G. Chiappino, A.S. Kelus, and P.G.H. Gell. 1965. Cellular localization of immunoglobulins with different allotypic specificities in rabbit lymphoid tissues. *J. Exp. Med.* **122:** 853.

Petras, J.L. 1972. An inherited albumin variation in the house mouse. *Biochem. Genet.* **7:** 1273.

Riggs, A.D. 1989. DNA methylation and cell memory. *Cell Biophys.* **15:** 1.

Rotello, R.J., M.G. Hockey, and L.E. Gerschenson. 1989. Biochemical evidence for programmed cell death in rabbit uterine epithelium. *Am. J. Pathol.* **3:** 491.

Schreiber, G., A. Tsykin, A.R. Aldred, T. Thomas, W.P. Fung, P.W. Dickson, T. Cole, H. Birch, F.A. De Jong, and J. Milland. 1989. The acute phase response in the rodent. *Ann. N.Y. Acad. Sci.* **557:** 61.

Smith, J.R. and R.G. Whitney. 1980. Intraclonal variation in proliferative potential of human diploid fibroblasts: Stochastic mechanism for cellular aging. *Science* **207:** 82.

Swindale, N.U. 1980. A model for the formation of occular dominance stripes. *Proc. R. Soc. Lond. B Biol. Sci.* **208:** 243.

Theiler, K. 1989. *The house mouse: Atlas of embryonic development.* Springer-Verlag, New York.

Till, J.E. 1981. Cellular diversity in the blood-forming system. *Am. Sci.* **69:** 522.

Till, J.E. and E.A. McCulloch. 1980. Hemopoietic stem cell differentiation. *Biochim. Biophys. Acta* **605:** 431.

Tomei, L.D., P. Kanter, and C.D. Wenner. 1988. Inhibition of radiation-induced apoptosis in vitro by tumor promoters. *Biochem. Biophys. Res. Commun.* **30:** 155.

Trauth, B.C., C. Klas, A.M. Peters, S. Matzku, P. Mololer, W. Falk, K.M. Debatin, and P.H. Krammer. 1989. Monoclonal antibody-mediated tumor regression by induction of apoptosis. *Science* **245:** 301.

Wadewitz, A.B. and R.A. Lockshin. 1988. Programmed cell death: Dying cells synthesize a co-ordinated, unique set of proteins in two different episodes of cell death. *FEBS Lett.* **241:** 19.

Programmed Cell Death and Apoptosis

R.A. Lockshin[1] and Z. Zakeri [2]

[1]Department of Biological Sciences, St. John's University
Jamaica, New York 11439
[2]Department of Biology, Queens College and the
Graduate Center of CUNY, Flushing, New York 11367-1597

The concepts of programmed cell death and apoptosis arose at approximately the same time and, without advocacy or specific intent, have been considered to be related or even interchangeable. The assumption of synonymy is an overextension of the relationship, and this book provides an opportunity to attempt to draw some distinctions and to reestablish the differences between the two phenomena.

There is very little doubt today that cells can die in a controlled manner. This sense has been implicit in the historical recognition of embryonic, metamorphic, and hormonoprivic cell death and cell turnover, with the controlled nature of the latter being most recently championed by Wyllie and others (Wyllie et al. 1990; R. Schulte-Hermann and W. Bursch, in prep.; Kerr and Harmon, this volume). The fact that none of these losses provoked inflammation or necrosis was acknowledged by many researchers and finally highlighted by Kerr et al. (1972). These latter investigators emphasized the substantially different morphology of what they termed apoptosis from classic necrosis, and they further contributed to the field in recognizing that mild pathological influences could evoke similar changes.

Programmed Cell Death

The existence of cell death in embryos has been recognized since the 19th century, when embryologists recognized the basis of sexual differentiation, formation of complex structures such as limbs, and disappearance of evolutionarily vestigial organs. Elaborate studies of the destruction of larval tissues in insects and amphibia proliferated in the first quarter of the 20th century. Examination of the loss of neurons led Hamburger, Levi-Montalcini, and Cohen to the discovery of nerve growth factor (Levi-Montalcini 1966) and led Glücksmann (1951) to categorize types of cell death according to evolutionary significance of the event. Saunders and his colleagues were the first to approach the question in an experimental manner (Saunders 1966; Saunders and Fallon 1966). They identified a region on the chick posterior necrotic zone (PNZ) in which substantial cell loss helped to define the border of the wing bud. Cell death in the region was identified by nile-blue-sulfate uptake: that is, by recognition of phagocytosed cells in phagocytic vacuoles. They first explanted the region as soon as it could be identified and grew it in culture, observing that at the appropriate age, numerous cells died in the culture. More interestingly, if they transplanted equivalent tissue to the back of the chick embryo, the tissue healed in place and survived. They interpreted this observation to mean that, although the cells carried within themselves the information leading to their death, since they would involute in the explant, they were not irreversibly damaged, since they could survive in another environment. In Saunder's phrase, "The death clock was ticking."

At approximately the same period, Lockshin and Williams, who were in contact with Saunders, observed a similar phenomenon in insect metamorphosis. In Saturniid silk moths, the intersegmental muscles degenerated after the ecdysis of the moth. Although several biochemical changes premonitory of death, such as increase in lysosomal enzymes, could be identified, as could endocrine and neural timing events, death of the muscles could be blocked by any of several neuropharmacological agents. Thus, it appeared to these authors that there existed a program or sequence of events leading to involution, and the term "programmed cell death" was born (Lockshin and Williams 1964, 1965a,b,c,d).

Inherent in both the Saunders and Lockshin/Williams images was the concept that the cell carried within itself the capacity to commit suicide when requested to by the organism, and that this gesture satisfied an evolutionary imperative to nourish or otherwise maintain the organism. Thus, comparisons were usually drawn to terminal differentiation of epidermis or mammalian erythrocytes. Although not articulated by either group, the presumption was that the sequence was genetic in origin but promoted by unknown means. Saunders, for instance, examined cell death in web-footed avians (Saunders and Fallon 1966), and Lockshin surveyed homologous tissues in other species. Bibliographies therefore contained references also to a small catalog of embryonic abnormalities characterized by variation in the amount of cell death. The intrinsic origin of the death was also presumed from microscopic images: Except for the very highly evolved Hymenoptera, there was no evidence for attack of otherwise healthy cells by phagocytes or noxious agents. The same image underlies embryonic neuronal cell death as well as cell death in *Caenorhabditis*, for which specific genes controlling specific cell deaths have been identified (Ellis and Horvitz 1966).

Two characteristics of programmed cell death were thus defined: (1) It is a developmental event; (2) it arises as a cellular response to an unknown stimulus. These observations, of course, led to a search for evidence for activation of genes, culminating in the late 1960s and early 1970s with the recognition that inhibition of protein synthesis can delay or prevent cell death (Tata 1966; Lockshin 1969; Munck 1971; Martin et al. 1988; Ucker et al. 1989; Oppenheim et al. 1990).

Early literature on apoptosis emphasized the generality of the phenomenon and cited such examples as involution of tadpole tail. For programmed cell death, the nuclear pycnosis and condensation of the cytoplasm resembled the margination of chromatin and shrinkage or delamination of the cytoplasm in apoptosis more than necrosis, and (particularly with evidence from amphibian metamorphosis) most researchers concluded that programmed cell death was apoptotic in nature.

Although this latter generalization may be valid, most well-documented forms of programmed cell death differ in characteristic ways from classic apoptosis, and casual melding of the two concepts leads to substantial confusion. Most importantly, *programmed cell death is an operational definition of a functional relationship in development, whereas apoptosis is an originally morphological, now biochemical description of a mode of cell death.* It is fruitful and stimulating to assume that there are very few means by which cells die naturally and that these means are subsumed into the rubric of, or are funneled into the final common pathway of, apoptosis; but the concept of apoptosis does not necessarily include programming.

Is Programmed Cell Death Apoptotic in Nature?

Cells dying for developmental reasons resemble an apoptotic lymphocyte far more than they resemble a necrotic cell, most particularly showing controlled condensation of the cell, lack of inflammatory response, and condensation of the chromatin. However, the morphology differs in several notable respects. Programmed cells frequently manifest involvement of lysosomes, and they rarely evidence laddering of the DNA (see Table 1). For the moment, it seems fruitful to consider programmed cell death to be a type of apoptotic (controlled) cell death, recognizing that the sequence of events may differ according to whether the death is planned in a developmental sequence or is a potential solution to an inadvertent but dangerous situation. A major question is whether the funnel into a final common pathway is so wide and short as to be uninteresting, but, in order to make the assessment, one must assume that the funnel exists.

Secondary Nature of the Endonuclease

There is very little evidence that the DNA endonuclease (calcium-activated endonuclease, CAN) activated in thymocyte apoptosis is activated early in other forms of programmed cell death. A suspension of thymocytes, which in any case contain little cytoplasm, provides an extremely accessible means of collecting fragments of DNA. Tissues bound by basement membranes or requiring homogenization provide opportunities at several steps at which DNA fragments might be lost. In the most clear-cut case for a secondary effect of the endonuclease (Ellis and Horvitz 1986), genetic loss of an endonuclease gene does not block cell death, but the nuclei persist far longer than normally. In other instances of unequivocally programmed cell death, such as metamorphosis of the insect intersegmental

TABLE 1 *CHARACTERISTICS OF PROGRAMMED CELL DEATH, APOPTOSIS, AND NECROSIS*

Characteristic	Programmed cell death	Apoptosis	Necrosis
Morphology	condensation of cell, fragmentation	condensation, fragmentation	lysis
Membrane integrity	persists until late	persists until late	early failure
Mitochondria	often, specific wave of autophagocytosis	unaffected	swelling, Ca^{++} uptake
Chromatin	condensed, electron-dense	margination	pycnosis
Autophagy	often, not invariably	absent	absent
Occult phase	several hours	minutes to hours	none
Protein synthesis	death blocked by actinomycin D, cycloheximide	death sometimes blocked by actinomycin D, cycloheximide	not affected by antibiotics
Origin	embryonic development, metamorphosis	glucocorticoids/lymphatic cells, ↓ trophic hormones or hypertrophic stimulus, mild toxic stimulus	strong toxin, hypoxia, large shift in pH
Paradigmatic cell	embryonic neuron, sexual or limb; interdigital tissue, insect; intersegmental muscle, labial gland	thymocyte + glucocorticoid, postcastration prostate	liver + hepatotoxins
Proximate control	endocrine, local or undefined, some neuroendocrine	↓ trophic hormones or stimuli	toxins
Cytoplasmic biochemical change	sometimes ↑ lysosomal enzymes, rarely heat shock proteins, c-myc, c-fos	no ↑ lysosomes; c-myc, c-fos?	rupture of lysosomes, no synthesis
Nuclear biochemical change	no laddering of DNA	laddering (internucleosome cleavage)	diffuse degradation
First manifestation	↓ protein synthesis, change in pattern	activation of endonuclease	blebbing or swelling

(↑) Increased; (↓) decreased.

muscles and labial gland, chromatin manifestly compacts but does not marginate in the classic thymocyte pattern (Beaulaton and Lockshin 1977). In the involuting labial gland, in which 100% of the cells die, electrophoresis of DNA fails to demonstrate nucleosomal ladders, whether the DNA is stained with ethidium bromide or prelabeled with tritium (R.A. Lockshin and N.M. Kodaman, in prep.). Furthermore, in the labial gland, the rate of protein synthesis drops at least 48 hours, and respiration begins to decrease 24 or more hours, before nuclear or other morphological changes are seen (L. Yesner and R.A. Lockshin, in prep.). Similarly, Zakeri et al. (1990) failed to identify nucleosomal ladders in the programmed cell death of the interdigital region of mouse limbs during both normal development and retinoic-acid-induced expanded cell death. There are many reasons to doubt these negative results, including both presumptive low frequency of dying cells in a tissue and loss of DNA fragments during isolation. The fact that fragments are not easily seen, however, suggests that perhaps endonucleases are not rate-limiting.

Requirement for Protein Synthesis

A requirement for protein synthesis has been documented for programmed cell death in numerous developmental situations. In the 1960s, the antimetabolites actinomycin D and cycloheximide were shown to block cell death in insect intersegmental muscle (Lockshin 1969; Lockshin and Beaulaton 1974b) and tadpole tail (Tata 1966), as well as in glucocorticoid-induced involution of thymocytes (which may not be programmed cell death) (Munck 1971). More recently, a similar requirement has been demonstrated for embryonic sympathetic neurons (Oppenheim et al. 1990). Apoptosis, on the other hand, has been shown to occur in the absence of protein synthesis and, in some circumstances, may even be induced by blockage of protein synthesis (Goldman et al. 1983).

There are many alternative reasons that antimetabolites might prevent cell death. As Carson has argued, if a cell's energy resources are restricted and protein synthesis represents a substantial drain on the energy, blockage of synthesis would prolong the life of the cell (Carson et al. 1986, 1988). For in vivo experiments, a paradigm of unintended effects is Szego's demonstration in the 1960s that actinomycin D inhibited responses to estradiol, but not if the animal was adrenalectomized; in other words, the blockage of response to estradiol was by glucocorticoids following administration of a highly toxic drug (Lippe and Szego 1965). Thus, the concept of requirement for protein synthesis has always at best been subject to challenge.

More convincing are the recent findings from several laboratories that the pattern of transcription changes markedly in dying cells. Wadewitz and Lockshin (1988) reported that, in *Manduca sexta*, collapse of the intersegmental muscles is presaged by disappearance of synthesis of routine messages such as for the myofibrillar proteins. Such a response is to be expected and has recently been confirmed (Schwartz et al. 1990). Of greater interest are the transcripts that are up-regulated as the cells die. Wadewitz and Lockshin (1988) and Lockshin and Wadewitz (1990) recognized several that were reproducibly up-regulated in two waves of cell death in homologous muscles. These transcripts appeared to be reasonably stable, since they could be identified during the several hours preceding and following the depolarization of the muscles. They did not identify any of the transcripts, which were characterized by their ability to direct the synthesis of proteins recognized only by their mobility on two-dimensional gels. L. Yesner (in prep.) found a similar conversion of pattern of protein synthesis during the involution of the labial gland but was unable to confirm the existence of proteins common to involuting labial glands and muscle. Yesner documented the absence of all heat shock proteins (hsp) except the ecdysone-induced hsp24. Schwartz et al. (1990), following a similar direction, confirmed the loss of myofibrillar message and isolated four clones of up-regulated genes. One of these proved to be ubiquitin, which the latter authors consider not to be required for cell death. However, a clone for a high-molecular-weight protein appeared to be temporally associated with the onset of involution of the muscle.

Other laboratories are seeking up-regulated genes in degenerating sympathetic neurons (Martin et al. 1988) and testosterone-deprived prostate epithelium. In the latter, genes are up-regulated, including c-*myc*, c-*fos*, and hsp70 (Buttyan et al. 1988), and a message for a protein of unknown function known variously as TRPM-2, clusterin, or sulfated glycoprotein-2. TRPM-2 is a curious protein, since it appears to be a secreted protein that is natively low in prostate and high in epididymis (Cheng et al. 1988, 1990; Grima et al. 1990). It also is up-regulated in several other instances in which inflammatory necrosis may be occurring, as well as in tissues that are differentiating in the absence of cell death. At present, as discussed in other chapters, the role and significance of TRPM-2 in cell death are unknown.

These conundrums raise the question of the extent to which any up-regulated gene is reactive or causal in the sequence leading to programmed cell death. Furthermore, the breadth of the vision of apoptosis is not clear: It may be very important for thymocytes to destroy DNA quickly, and therefore to activate an endonuclease, but cleavage of DNA appears to be a more secondary event in many forms of programmed cell death (Lockshin and Beaulaton 1974a; Beaulaton and Lockshin 1977, 1982; Beaulaton et al. 1986; Ellis and Horvitz 1986; Masters et al. 1988.) This argument has been most directly addressed by Horvitz's group, who have exploited the readily manipulated genetics of *Caenorhabditis elegans* to identify a sequence of genes required for cell death (Ellis and Horvitz 1986; Yuan and Horvitz 1990). The identities of these genes have not yet been published. Whether in this system or others, proof of the contention that specific genes control cell death will come from direct control of the activities of the genes by genetic or molecular means.

The examples cited above refer strictly to situations of normal development. Similar efforts to identify up-regulated genes have examined glucocorticoid-induced involution of thymocytes and pathological apoptosis in liver. Young (1984) recognized a few changes in proteins in thymocytes, when he examined dying and surviving thymocytes by high-resolution two-dimensional electrophoresis, and transglutaminase has been seen to be up-regulated in liver pathology (Fesus et al. 1987; Fesus and Thomazy 1988). The synonymy of apoptosis, as opposed to programmed cell death, with a requirement for protein synthesis is far more dubious. Depending on circumstances leading to unscheduled or unprogrammed apoptotic cell death (toxin, deprivation of trophic hormone or growth factor, glucocorticoid suppression), protein synthesis may not precede involution.

What Proteins Are Synthesized in Programmed Cell Death?

In evaluating the question of the breadth of the funnel leading to the final common pathway, one could speculate on several potential roles for required protein synthesis:

1. Transcriptional systems could be generally or specifically inhibited, forcing the cell to collapse as its most vulnerable component disappears. There is some evidence for this argument, since the earliest change in labial glands is a sharp decrease in overall protein synthesis.
2. Specific lytic enzymes could be induced to destroy individual components of cells. Although certainly lysosomal enzymes increase in many forms of cell death, there is little support for the argument that these enzymes cause cell death rather than react to a deteriorating situation.
3. Energy resources could be severely compromised by synthesis of inhibitors or blockage of synthesis of critical components. Although there is some evidence that such a situation may obtain in some cases—for instance, the intersegmental muscles become hypoxic and mitochondria are removed in a wave of apoptosis—the finding appears to be neither universal nor dramatic.
4. Specific destruction of vital components may be lethal to the cell. Although this argument is intellectually appealing, there is currently no evidence that membranolytic or other factors compromise the integrity of the cell until quite late in the process of involution. In those instances that have been studied, cells neither depolarize nor become permeable to larger molecules during the earlier phases.

Apoptosis vs. Programmed Cell Death: Which Factors Are Common, and Which Are Local to the Individual Cell?

It is not clear whether programmed cell death distinctly differs from unscheduled apoptosis or whether the differences seen merely reflect differences in cell priorities. A thymocyte is a potentially highly mitotic cell on the front line of defense against agents readily capable of damaging or mutating its DNA, and furthermore, it contains only a token amount of cytoplasm. Thus, there are many reasons to presume that rapid destruction of its DNA is a highly desirable evolutionary decision. Insect larval muscle, in contrast, is a postmitotic tissue that consists primarily of structural protein; in this situation the evolutionary imperative would be quite different. Thus, while the decision to shut down a cell might be achieved via a common pathway, the manifestations might be substantially different. In fact, we know very little about the mechanism of cell death.

How Does the Cell Know When to Die?

The question of whether programmed cell death and apoptosis are synonymous cannot be resolved until we understand the sequence of signals that lead a cell to collapse. Currently we know that various biological molecules can either evoke involution or protect against it and that numerous mildly but generally toxic agents can also trigger apoptosis. We know virtually nothing of the steps between the impinging influence and the beginning of collapse. Earlier research established that depletion of energy from glucocorticoid-treated thymocytes neither anticipated nor caused apoptosis, and in other systems, loss of other phosphorylated intermediates may be important (Carson et al. 1986, 1988). Ultimately in apoptosis, CAN is activated and destroys the DNA, coincident with (perhaps physical) alteration of the cell membrane; in programmed cell death, a series of genes of unknown function and consequence are activated. In the chick wing, an early sign of programmed cell death is the withdrawal of cells from mitosis; in the postmitotic insect model, change in pattern of synthesis of proteins and perhaps substantial decrement in total protein synthesis are premonitory. If programmed cell death is apoptotic, then presumptively the change in synthetic pattern, like other responses to hormones or growth factors, engenders a series of metabolic changes in the cell. In this instance, however, these changes coalesce to create a situation akin to that created less purposefully by toxins or more general stresses, and a collapse similar to apoptosis ensues. All or most cells are capable of undergoing apoptosis, and the programming of death is a sequence that leads ineluctably to force the cell into apoptosis. By this interpretation, the morphology of collapse would vary from the classic image of apoptosis according to the biology of the cell.

What is lacking from this image is a clear description of the steps between the reception of the signal and the initiation of collapse. These are potentially the most interesting, because they are capable both of giving us an understanding of the cell's perception of its status and of allowing us to homologize or differentiate between programmed cell death and apoptosis. Our responsibility to this research field is to analyze the unknown steps so that we can follow the complete sequence of programmed cell death from the initial call to self-destruct to final removal of the expired cell.

ACKNOWLEDGMENT

Research reported here was partially supported by a grant from the March of Dimes (Z.Z.).

REFERENCES

Arends, M.J., R.G. Morris, and A.H. Wyllie. 1990. Apoptosis: The role of the endonuclease. *Am. J. Pathol.* **136**: 593.

Beaulaton, J. and R.A. Lockshin. 1977. Ultrastructural study of the normal degeneration of the intersegmental muscles of *Antheraea polyphemus* and *Manduca sexta* (Insecta, Lepidoptera, with particular reference to cellular autophagy. *J. Morphol.* **154**: 39.

Beaulaton, J. and R.A. Lockshin. 1982. The relation of programmed cell death to development and reproduction: Comparative studies and an attempt at classification. *Int. Rev. Cytol.* **29**: 215.

Beaulaton, J., G. Nicaise, G. Nicolas, and R.A. Lockshin. 1986. Programmed cell death. X-ray microanalysis of calcium and zinc within the electron-dense droplets of the T system in insect muscles. *Biol. Cell.* **56**: 271.

Buttyan, R., Z. Zakeri, R.A. Lockshin, and D. Wolgemuth. 1988. Cascade induction of c-*fos*, c-*myc*, and heat shock 70 k transcripts during regression of the rat ventral prostate gland. *Mol. Endocrinol.* **2**: 650.

Carson, D.A., C.J. Carrera, D.B. Wasson, and H. Yamanaka. 1988. Programmed cell death and adenine deoxynucleotide metabolism in human lymphocytes. *Adv. Enzyme Regul.* **27**: 395.

Carson, D.A., S. Seto, D.B. Wasson, and C.J. Carrera. 1986. DNA strand breaks, NAD metabolism, and programmed cell death. *Exp. Cell Res.* **164**: 273.

Cheng, C.Y., C.-L.C. Chen, Z.M. Feng, A. Marshall, and C.W. Bardin. 1988. Rat clusterin isolated from Sertoli-cell enriched culture medium is sulfated glycoprotein-2 (SGP-2). *Biochem. Biophys. Res. Comm.* **155**: 398.

Cheng C.Y., J. Grima, M.S. Stahler, R.A. Lockshin, and C.W. Bardin. 1990. Testins are structurally related sertoli cell proteins whose secretion is tightly coupled to the presence of germ cells. *J. Biol. Chem.* **264**: 21386.

Ellis, H.M. and H.R. Horvitz. 1986. Genetic control of programmed cell death in the nematode *C. elegans*. *Cell* **44**: 817.

Fesus, L. and V. Thomazy. 1988. Searching for the function of tissue transglutaminase: Its possible involvement in the biochemical pathway of programmed cell death. *Adv. Exp. Med. Biol.* **231**: 119.

Fesus, L., V. Thomazy, and L. Falus. 1987. Induction and activation of tissue transglutaminase during programmed cell death. *FEBS Lett.* **224**: 104.

Glücksmann, A. 1951. Cell deaths in normal vertebrate ontogeny. *Biol. Rev. Camb. Philos. Soc.* **26**: 59.

Goldman, A.S., M.K. Baker, R. Piddington, and R. Herold. 1983. Inhibition of programmed cell death in mouse embryonic palate in vitro by cortisol and phenytoin: Receptor involvement and requirement of protein synthesis. *Proc. Soc. Exp. Biol. Med.* **174**: 239.

Grima J., I. Zwain, R.A. Lockshin, C.W. Bardin, and C.Y. Cheng. 1990. Diverse secretory patterns of clusterin by epididymis and prostate/seminal vesicles undergoing cell regression following orchiectomy. *Endocrinology* **126**: 2989.

Kerr, J.F.R., A.H. Wyllie, and A.R. Currie. 1972. Apoptosis: A basic biological phenomenon with wide-ranging implications in tissue kinetics. *Br. J. Cancer* **26**: 239.

Levi-Montalcini, R. 1966. The nerve growth factor: Its mode of action on sensory and sympathetic nerve cells. *Harvey Lect. Ser.* **60**: 217.

Lippe, B.M. and C.M. Szego. 1965. Participation of adrenocortical hyperactivity in the suppressive effect of systemic actinomycin D on uterine stimulation by oestrogen. *Nature* **207**: 272.

Lockshin, R.A. 1969. Programmed cell death. Activation of lysis of a mechanism involving the synthesis of protein. *J. Insect Physiol.* **15**: 1505.

Lockshin, R.A. and J. Beaulaton. 1974a. Programmed cell death. Cytochemical evidence for lysosomes during the normal breakdown of the intersegmental muscles. *J. Ultrastruct. Res.* **46**: 43.

———. 1974b. Programmed cell death. Cytochemical appearance of lysosomes when the death of the intersegmental muscles is prevented. *J. Ultrastruct. Res.* **46**: 63.

Lockshin, R.A. and A.G. Wadewitz. 1990. Degeneration of myofibrillar proteins during programmed cell death in *Manduca sexta*. *UCLA Symp. Mol. Cell Biol.* **123**: 283.

Lockshin, R.A. and C.M. Williams. 1964. Programmed cell death. II. Endocrine potentiation of the breakdown of the intersegmental muscles of silkmoths. *J. Insect Physiol.* **10**: 643.

————. 1965a. Programmed cell death. I. Cytology of degeneration in the intersegmental muscles of the Pernyi silkmoth. *J. Insect Physiol.* **11**: 123.

————. 1965b. Programmed cell death. III. Neural control of the breakdown of the intersegmental muscles of silkmoths. *J. Insect Physiol.* **11**: 605.

————. 1965c. Programmed cell death. IV. The influence of drugs on the breakdown of the intersegmental muscles of silkmoths. *J. Insect Physiol.* **11**: 803.

————. 1965d. Programmed cell death. V. Cytolytic enzymes in relation to the breakdown of the intersegmental muscles of silkmoths. *J. Insect Physiol.* **11**: 831.

Martin, D.P., R.E. Schmidt, P. DiStefano, O. Lowry, J. Carter, and E. Johnson. 1988. Inhibitors of protein synthesis and RNA synthesis prevent neuronal death caused by nerve growth factor deprivation. *J. Cell Biol.* **106**: 829.

Masters, J.N., C.E. Finch, and R.M. Sapolsky. 1988. Glucocorticoid endangerment of hippocampal neurons does not involve deoxyribonucleic acid cleavage. *Endocrinology* **124**: 3083.

Munck, A. 1971. Glucocorticoid inhibition of glucose uptake by peripheral tissues: Old and new evidence, molecular mechanisms, and physiological significance. *Perspect. Biol. Med.* **14**: 265.

Oppenheim, R.W., D. Prevette, M. Tytell, and S. Homma. 1990. Naturally occurring and induced neuronal death in the chick embryo in vivo requires protein and RNA synthesis: Evidence for the role of cell death genes. *Dev. Biol.* **138**: 104.

Saunders, J.W., Jr. 1966. Death in embryonic systems. *Science* **154**: 604.

Saunders, J.W. and J.F. Fallon. 1966. Cell death in morphogenesis. In *Major problems in developmental biology*, (25th Symposium of the Society for Developmental Biology) (ed. M. Locke), p. 289. Academic Press, New York.

Schwartz, L.M., L. Kosz, and B.K. Kay. 1990. Gene activation is required for developmentally programmed cell death. *Proc. Natl. Acad. Sci.* **87**: 6594.

Tata, J.R. 1966. Requirement for RNA and protein synthesis for induced regression of tadpole tail in organ culture. *Dev. Biol.* **13**: 77.

Ucker, D.S., J.D. Ashwell, and G. Nickas. 1989. Activation-driven T cell death: I. Requirements for de novo transcription and translation and association with genome fragmentation. *J. Immunol.* **143:** 3461.

Wadewitz, A. and R.A. Lockshin. 1988. Programmed cell death. Dying cells synthesize a coordinated unique set of proteins in two different episodes of cell death. *FEBS Lett.* **241:** 19.

Young, D.A. 1984. Advantages of separations on "giant" two-dimensional gels for detection of physiologically relevant changes in the expression of protein gene-products. *Clin. Chem.* **30:** 2104.

Yuan, J. and H.R. Horvitz. 1990. The *Caenorhabditis elegans* genes ced-3 and ced-4 act cell autonomously to cause programmed cell death. *Dev. Biol.* **138:** 33.

Zakeri, Z., A. Alles, P. Xia, L. Yesner, N. Cardamon, and R.A. Lockshin. 1990. Difficulty of recognizing, and potential lack of, early DNA degradation in programmed cell death in mice and insects. *J. Cell Biol.* **111:** 494a.

Carcinogenesis and Apoptosis: Paradigms and Paradoxes in Cell Cycle and Differentiation

F.O. Cope[1] and J.J. Wille[2]
[1]Medical Department, Ross Laboratories
Columbus, Ohio 43215
[2]Skin Physiology Program, ConvaTec
Division of Bristol-Myers Squibb Company
Skillman, New Jersey 08558

In this paper, we attempt to place in perspective the concepts of programmatic death (*apoptosis*) and aspects of carcinogenesis. We recommend that the reader acquire historical and scientific information on "programmed" cell death prior to evaluating this chapter by scanning the chapters by Kerr and Harmon and Lockshin and Zakeri (both this volume).

Our views are, in all probability, in contrast to those of our colleagues in this volume, as there are many concepts of what apoptosis is and what is or should be included in such a concept. However, we all agree that the concept of apoptosis is having and will continue to have a profound effect on thinking in cell biology and adherent research areas, even to the degree of significantly altering paradigms and testing the most rigid dogma. Indeed, this is the purpose of this volume and of this chapter about cellular processes.

Any endeavor to maintain life must in some respects be programmatic. Most such programs are, organismally speaking, deterministic. However, as we stratify our knowledge about the organization of cell processes, we are increasingly faced with the fact of creeping stochasticism. (For additional discussions, see Michaelson; Tomei, both this volume.) This distillation of our queries then must be pragmatic and lead us to a quantitative investigation of how cell processes are modu-

Apoptosis: The Molecular Basis of Cell Death
Copyright 1991 Cold Spring Harbor Laboratory Press 0-87969-366-5/91 $3.00 + 00

lated in order to maintain a viable transduction entity, the organism. To address the role of apoptosis in cell biology and its particular relationship to carcinogenesis and anticarcinogenesis, cell cycle, and differentiation, we incorporate several dogmas (as limited in and of themselves as they may be) in the discussion: (1) multistage carcinogenesis, including *initiation* and *promotion*; (2) oncogenes; (3) metastasis; (4) cell signaling; (5) cell cycle (mitosis); and (6) differentiation. We attempt to separate each of these relative to the ability and the potential for any cell to undergo apoptosis or to respond to the proximal or distal expression of apoptosis.

Apoptosis Versus Carcinogenesis Versus Anticarcinogenesis

Obviously, the organism, organ systems, tissues, and cells are profound transducing agents, each a stratum in and of its own right, as well as interdependent. Even more profound and paradoxical is the ability of any malignant cell or tumor to be an exceptional transducing system, not only usurping proximal and distal signals, but also producing its own signals and "manufacturing" its own adaptation. What then are the key elements separating the normal process from the malignant one, and why does a process such as aging lead to increased risk of expression of the malignant phenotype? The intuitive response is, like Hayflick's concept of limited cell division potential (Hayflick 1961, 1965), that aging and its constituents including malignant risk are manifest "programmed cell death" with cumulative programmatic losses and increased error-proneness (Goldstein 1990). Thus, we are required at this point to define exactly what the major default focus is relative to segregating aging, carcinogenesis, and apoptosis and their operatives, cell cycle, and differentiation (signaling and gene-cassette programming). These operatives represent temporal and spatial aspects, respectively. For the record, we state that the focus, both in an experimental sense and in a hypothetical sense, is *fidelity*.

The mechanisms of biological responses studied in vitro are likely to involve apoptosis in some manner. For example, multistage carcinogenesis is a widely accepted model for a malignant transformation (fidelity, if you will), but the problem of integrating temporal and spatial events between the *initiating* events and expression of the transformed phenotypes through the *promotion* and *progression* processes is profound. Presumably, initiation involves the introduction of specific DNA lesions, and there is excellent correlation between the properties of carcinogens and mutagens (Ladik 1985; Takeshita et al. 1985). Additionally, initiation, followed by the biologically coordinate events of promotion, can lead to the expression of the malignant phenotype. Although numerous hypotheses have been made regarding requisite somatic changes in carcinogenesis, there is little doubt that the induction of direct or indirect DNA damage is important. Thus, what is introduced into this particular process is the diminution of the cell's fidelity relative to expressing a proper dominant senescence program, integrating a proper senescence program, or responding to proximal or distal signals, all of which, in some fractional population sense, contain an apoptotic component (Pereira-Smith and Smith 1988; Sugawara et al. 1990). Compounding this particular concept is the recent focus on the expression and roles of *oncogenes*. The question in this case is not whether oncogenes themselves cause cancer, or lead to a malignant phenotype, but rather how the role of oncogenes is assessed relative to their temporal and spatial expression during the diminution of fidelity or spatial shifts, even temporary ones, in genomic "cassette" programming; indeed, oncogenes can be construed to be a manifest reduction of programmatic fidelity in any cell or cell population where two expression strata are asynchronous, for example, senescence versus immortalization (see below). The demonstration that some tumor promoters inhibit apoptosis and induce the expression of several oncogenes suggests that the role for these compounds focuses on resetting of genomic cassette programs leading to de novo gene expression, altered integration of epigenetic events, and malignant transformation.

Added to this notion, and segregating apoptosis from tumor promotion, is the observation that apoptosis-inducing agents do not induce mitosis but that both may be induced simultaneously (Table 1). More importantly, these data suggest that mitosis and apoptosis exist as distinct genomic cassette strata, where common elements are not shared but where

TABLE 1   *RELATIVE INDUCTION OF MITOSIS AND APOPTOSIS BY TUMOR PROMOTERS AND ANTITUMOR PROMOTERS*

	Mitosis	Apoptosis
Phorbol esters (TP)	++++	++++
Okadaic acid (TP)	++++	0
Retinoic acid (ATP)	0	++++
Difluoromethyl ornithine (ATP)	0	+++

(TP) Tumor promoters; (ATP) Antitumor promoters.

common temporal elements may be shared. For example, retinoic acid (RA) blocks both the phorbol ester 12-*O*-tetradecanoylphorbol-13-acetate (TPA)- and okadaic acid (OA)-induced promotion (Cope 1989). TPA is a protein kinase C inducer whereas OA is not; OA is a suppressor of phosphatase 1 and 2A. Thus, although RA may induce a single cell-cycle phase or cassette shift, this is likely an "into phase" cassette shift leading to the induction of apoptosis and/or the acquisition of a new differentiation stratum with a competent senescence component containing an apoptotic element. Unfortunately, this shift may be temporally phased into a mitotically inauspicious program such that even though no elements are shared, they temporally complement one another, leading to the maintenance of the malignant phenotype! Such a concept can change the therapeutic approach to treating neoplastic disease where apoptotic realms are the focus as opposed to treating mitotic realms. We have shown this to be a preferred approach in human squamous cell carcinoma (Cope and Wille 1989).

Concomitant with this concept is the notion that oncogenes may be coexpressed and that they are part of an infidelity satellite gene cassette. Attacking apoptosis here is a new approach. Typically, mRNA is an important control point in the transient expression of a wide variety of oncogenes whose products are coincident with growth factors or their receptors (Travali et al. 1990; Tuck et al. 1991). Labile mRNAs of oncogenes contain within their 3'-untranslated region an AU-rich region shown to destabilize their messages (Gillis and Malter 1991; Malter and Hong 1991). Recently, a binding protein, adenosine-uridine-binding factor (AUBF), which complexes with four AUUUA tandem reiterations, was found; it stabilizes the proto-oncogene mRNAs. AUBF is categorically part of the mitotic gene motif, and it is induced by TPA (Malter and Hong 1991); this effect is antagonized by RA. Additionally, the *yin* and *yang* element of these antagonists is protein phosphorylation; TPA = up, RA = down, and phosphorylation-dependent gene expression appears to be an element that may be a focal area separating mitosis from apoptosis (Lenormand et al. 1990). Thus, we need to look at protein phosphorylation and other signal elements that define the mitotic and apoptotic gene realms. Several genes associated with cell death have been described in the nematode model, including the *ced-3* and *ced-4* genes (Yuan and Horvitz 1990). These genes are required for a total program of cell death, whereas in "deathless" mutants, *ced-1*, *ced-2*, and *Nuc-1* are predominantly expressed. Recently, Kyprianou et al. (1990) reported evidence from rat studies that cell death in the prostate following castration was associated with the induction of the TRPM-2 gene, and Slawin et al. (1990) observed that the induction of SGP-2 was associated with cell death in Sertoli cells. Such genes have been proposed as putative apoptosis genes, and the nature of gene products in the role of bringing about death probably relates to a programmatic in-phase collapse to terminal differentiation. Thus, like the normogenic pathway or the malignant pathway, apoptosis has specific genetic products that may be cell-specific and are instrumental in modulating the cell response within expedient temporal and spatial frameworks (Figs. 1–6). The momentary conclusion one draws from these observations is that apoptosis has little to do with a concept of programmed aging (loss of fidelity) and loss of organismal life. Indeed, one sees from these data that apoptosis is a process by which some cells undergo a program shift that positively influences proximal and distal cells and tissues in order to maintain local and distal program and signal response fidelity.

If any additional evidence is needed to support this particular notion, no better models exist than the cell-surface interactions described by Alnemri and Litwack (1990) or the involvement of programmed cell death in the regulation of embryonic development (Altaba and Melton 1989). Alnemri and Litwack

(1990) reported that specific cell-surface changes appeared in apoptotic target cells and were associated with enhanced recognition by cytotoxic cells. These cell-surface changes are accompanied by a release of soluble factors into the surrounding medium. Additionally, these products may enhance survival of other cells exposed to toxic stress, maintaining a viable pluripotent cell population. The control of cell death, therefore, must be considered not only in the view of a single cell, but more importantly, in terms of physiological control in entire cell populations, such as those in the developing embryo.

Embryogenesis and Apoptosis

In developing organisms, there is an establishment of axiopolarity and a patterning of cells that are dependent on local signals derived from particular regions of the embryo. What is most expedient in using embryogenesis or morphogenic development as an apoptosis cell model is the combining of observations relating to the function of RA. RA has been reported to be a true morphogen. Indeed, the review by Brockes (1991) suggests that RA and the expression of its nuclear receptors were key to controlling the formation of zones of polarizing activity in developing embryos. This suggests that RA is a centripetal molecule involved in pattern formation in different embryonic tissues and that signaling patterns are highly conserved, as well as the temporal and spatial expression of receptors that modulate the RA signal process. However, the ability of RA to form zones of polarizing activity has been questioned. Our work relating to the inhibition of specific RA receptors (α-nuclear RA receptor) suggests an alternative mechanism of action for retinoids, both in embryonic development and as this relates strongly to retinoids as antitumor promoter compounds (Cope and Wille 1989). The most recent work involving RA shows that it does not form a true zone of polarizing activity (Brockes 1991). Rather, RA controls some other spatial signaling process in the developing embryo. We put forth a hypothesis which also indicates that RA does not control the zone of polarizing activity and that the response of a human squamous cell carcinoma to the elimination of an RA receptor was akin to the process of RA involvement in the developing embryo. This suggests that in human squamous cell carcinoma, the elimination of the RA receptor-α induces the apoptosis program and that RA has the ability to induce a *metastable* apoptotic state in the embryo. In the malignant cell, when this metastable state is spatially and temporally coupled with a process that alters the mitotic clock, the malignant state is maintained. Our data are also consistent with the finding of L.D. Tomei (pers. comm.) that RA has the capability of blocking apoptosis in immortalized mesenchymal (10T1/2) cells exposed to irradiation.

A Further Cell Analysis

The question now is, How can we further reconcile our previous observations that tumor-promoting phorbol esters block apoptosis (see previous discussion) and that antitumor promoters such as RA also achieve this end point? The data involving the inhibition of apoptosis by tumor promoters such as TPA (Tomei et al. 1988), as well as by antitumor promoters such as RA, which also block apoptosis, suggest a very profound result. That is, gene-cassette programming as modulated by apoptosis is pseudocoupled to the mitotic clock. In the developing embryo, then, RA controls the relative signal response of cells in which receptors for RA are expressed while the mitotic index and the influence of density-dependent signaling per se are pseudocoupled to this RA signal. This signal consists of both a primary or epigenetic signal and a genetic signal, as we have previously shown in Chinese hamster lung cells (Cope 1989).

We conclude from these data that apoptosis is *not* related to the loss of fidelity; indeed, the contrary is true, that apoptosis is an adaptation related to the maintenance of fidelity of the collapse of cell programming to an isochronic focus, which may be senescence or terminal differentiation for *any* fractional population in the developing embryo and the maintenance of a functional phenotype in both tissue and organism. The RA data also suggest that although tumor promoters induce a bypass state relative to the induction of terminal differentiation, they also produce an extreme change in mitotic clock components. This is typically not true for antitumor promoters such as RA, which have a capacity to block apoptosis (Cope and Wille 1989). Thus, antitumor promoters relative to their ability to block apoptosis essentially are compounds which, rather than produce manifest changes in cell clock per se, produce changes that are programmatic cassette shifts, in pathways to terminal differentiation such as the differentiation of basal cells to fully differentiated epidermis or the switch of homeobox gene clusters in the developing embryo.

We sum up our view of the relative ability of any particular cell's undergoing a specific programmatic change relative to changes in mitotic clock potential in Figure 7. The four parameters are from left to right: log relative incidence of apoptosis; log mitotic potential; log relative response probability to an apoptotic signal; and relative probability of any cell to undergo apoptosis/10^6 cells. In malignancy, according to our model, cells have an extremely low potential in terms of incidence of apoptosis as well as relative probability of response to any apoptotic signal and a relatively low probability that any cell will undergo apoptosis; this is in contrast to mitotic potential. When contrasting this to some normal epithelial cell population, one can see that there is a significant increase in the incidence of apoptosis and that mitotic potential is significantly reduced, whereas response and the probability of undergoing apoptosis are relatively higher. In embryonic development relative to these other two states, there is an extremely high potential for all four states to exist concomitantly. However, this suggests that unlike the malignant state, the mitotic clock and apoptosis are not only pseudocoupled but that the mitotic clock is pseudostable relative to the changes in the program, falling into an eventual stable state. In initiation or promotion relative to the malignant state, these states tend to mimic the malignant state, although we hypothesize that the relative probability of any cell to undergo programmatic apoptosis is actually slightly greater in malignancy than it is during promotion. Where initiation is coupled to promotion, the overriding factor relative to formation of the malignant state is a resetting of the mitotic potential as induced by promotion, which becomes a *stable* state. When looking at an antipromoter such as RA relative to the tumor promoter TPA, one can see that although RA can significantly reduce the relative probability of any cell to enter apoptosis as well as significantly reduce the incidence of apoptosis, the mitotic potential is significantly reduced while the relative response probability to an apoptotic signal is maintained. This would be a preparative phase relative to a programmatic shift such as that occurring in the developing embryo subsequent to a release from this apoptotic state and the expression of new diffusible factors which either are part of the apoptotic program per se or are part of the program of proximal cells. In trauma, we indicate in our model that the relative incidence of apoptosis is significantly increased while the mitotic potential is only slightly increased. The increase in the incidence of apoptosis is also concomitant with a slight increase in the relative probability of any cell to undergo apoptosis and a slight reduction in the relative response probability to an apoptotic signal. This produces what we would describe as a preparatory or recovery state for any traumatic phase and is requisite for the maintenance of some pluripotent cell population post-trauma.

The way we have calculated these particular potentials is outlined in Equation 1.

Relative apoptotic program potential/10^6 cells =
$$\text{Log}_{10}\left[\{[(a) + (b)]/10^6 \text{ cells}\} \div (c)\right] \tag{1}$$

Here, we show that the relative apoptotic program potential/10^6 cells is equal to the $\log_{10}$ of the term where (a) is equal to the square of the number of usual apoptotic cells in that state plus (b) which is the number of program-receptive cells/10^6 cells. This entity is divided by (c), which is the epigenetic and genetic errors acquired during time per 10^9 bits of information (this quantity squared over repair capacity per 10^9 bits of information transferred). In our model, the complete denominator can be expressed as any fidelity function at time t or per unit time. In our model, repairs are assumed to be 100 bits per 10^9 bits which equal 100% fidelity. (See Table 2 for term values.)

Table 2 indicates the relative apoptotic program potential in each of the states outlined in Figure 7. This is a log-relative potential for normal, malignant, embryonic, initiation, promo-

TABLE 2 RELATIVE APOPTOTIC PROGRAM POTENTIAL

	(a)	(b)	(c)	Log relative potential
Normal	10^3	10^6	30	-0.6
Malignant	10	10^2	200	-6.3
Embryonic	5×10^4	10^6	24	0.8
Initiation	10^3	10^5	104	-1.9
Promotion	10	10^3	100	-4.9
Antipromotion	10	10^6	30	-1.0
Trauma	10^5	5×10^5	102	1.9

Values are logs and are estimated by substitution of mean values for each "unknown" in Equation 1.

tion, antipromotion, and trauma states. The log-relative potential for any normal cell to undergo an apoptotic program is approximately 1. Our model also indicates that there is at least a 500,000-fold difference in relative potential for a normal cell versus a malignant cell to undergo apoptosis. Conversely, in the embryonic state, the relative apoptotic program potential is tremendously increased, although the potential for a cell to undergo apoptosis during promotion is also very low. Relative to the malignant state, our hypothesis indicates that the sum of the log-relative potential of initiation or promotion is approximately equal to that of the malignant state, suggesting that their products relative to expression of changes in cell-cycle clocks and gene-cassette programming result in the malignant state.

What Does the Model Look Like?

As shown in Figure 1, there is an indication that in any particular population of cells, embryonic or epithelial cells, there is a pluripotent cell population that undergoes a process of normogenesis. These cells express temporal and spatial cell program decision points relative to being influenced by other cells or relative to their intrinsic program (cassette). As these cells move through the normogenic processes as indicated by the arrows (see y axis for relative pluripotency or morphogenic potential and x axis for time), there appears to be a stochastic

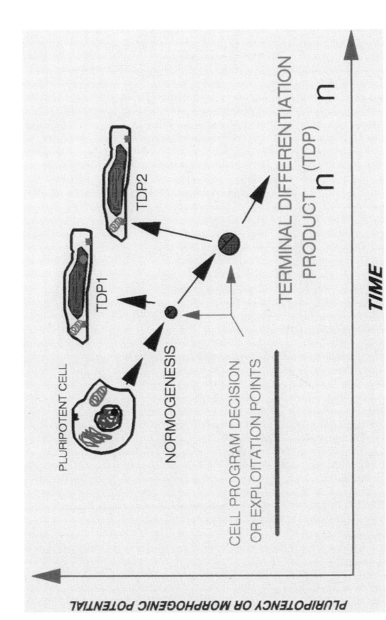

FIGURE 1 Interaction of gene-directed cell death-committed cells with pluripotent cell differentiation: Decision points.

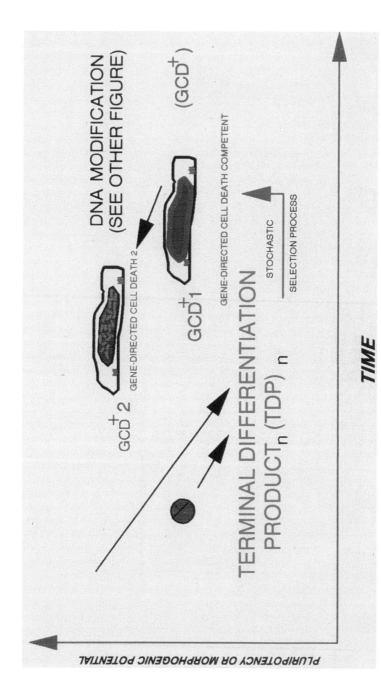

FIGURE 2 Interaction of gene-directed cell death-committed cells with pluripotent cell differentiation: Terminal differentiation product formation.

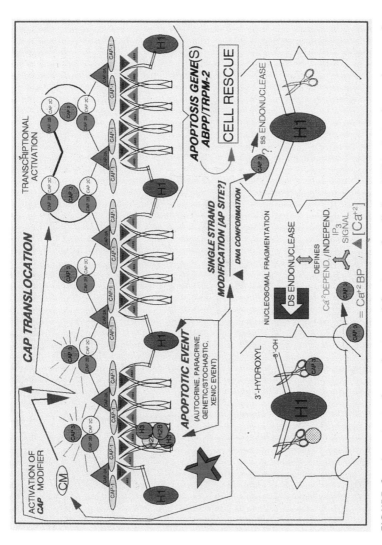

FIGURE 3 Apoptosis-dependent modification of cellular DNA: Interdependence of event initiation, chromatin-associated protein (CAP) strata, inositol phosphate signals, and the expression of amyloid-β-precursor protein (ABPP) or TRPM-2 in cell rescue.

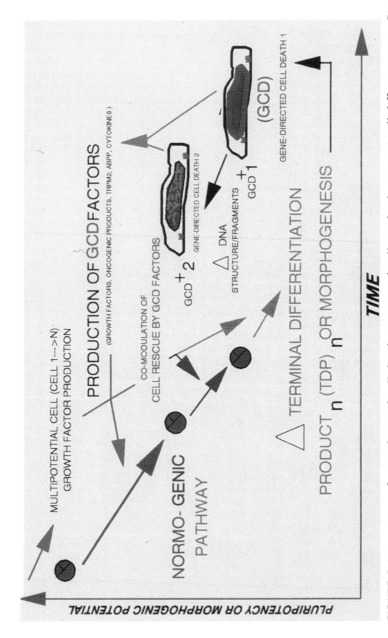

FIGURE 4 Interaction of gene-directed cell death-committed cells with pluripotent cell differentiation: Gene-directed cell death competency (GCD⁺) production of GCD factors.

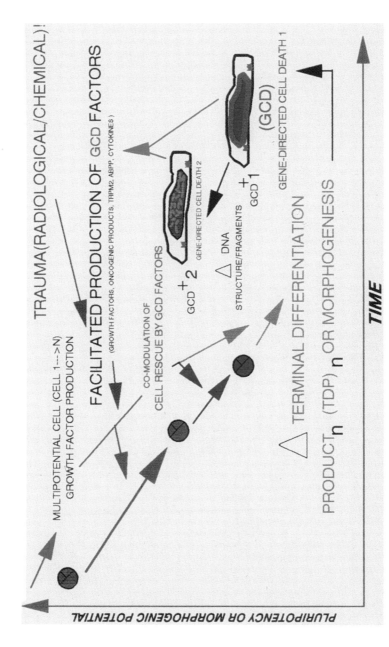

FIGURE 5 Interaction of gene-directed cell death-committed cells with pluripotent cell differentiation: Trauma, facilitated GCD factor production, and cell rescue.

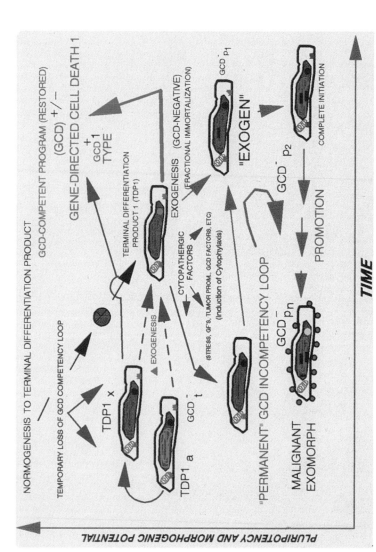

FIGURE 6 Escape from gene-directed cell death: Temporary outlooping (GCD_t^+) and "permanent" outlooping ($GCD_{p1}^- \rightarrow GCD_{pn}^-$).

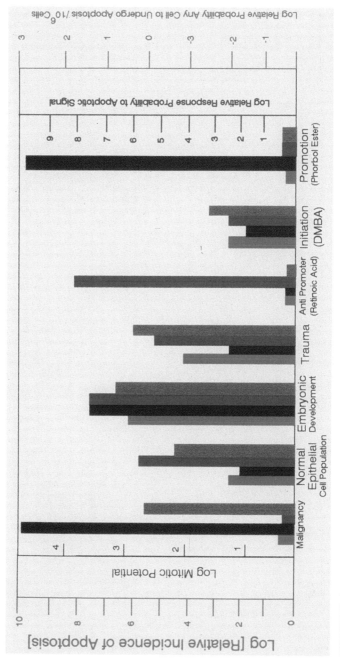

FIGURE 7 Apoptotic event probabilities, relative to modifications in cell programming.

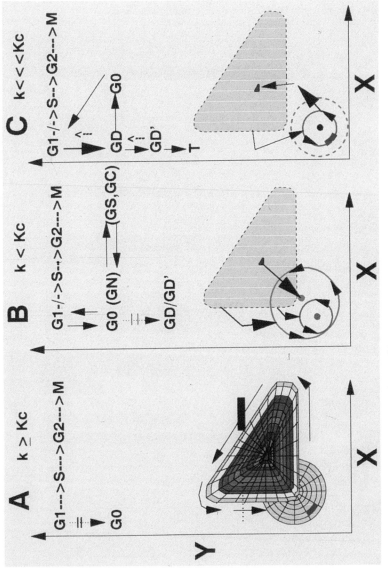

FIGURE 9 (See facing page for legend.)

process that leads to either a terminal differentiation product 1 or a terminal differentiation product 2 through terminal differentiation product n. At any terminal differentiation product (TDP$_n$; see Fig. 2), cells may undergo a process of apoptosis or gene-directed cell death; the observation that cells can undergo this change indicates that some subpopulation is GCD (gene-directed cell death)-competent (GCD$^+$). According to current observations, this process involves the margination of DNA and ultimately can result, in some cell types, in the fragmentation of DNA. This latter state would be the GCD$^+$2 state. The fact that DNA fragments in no way suggests that the cell has undergone "death," such that it was incapable of carrying out transcription and translation and the production of new cell products. Figure 4 indicates that concomitant with or subsequent to this particular process, apoptotic cells release growth factors and other products such as the TRPM-2 gene product discussed earlier.

The production of specific factors related to the expression of apoptosis in GCD$^+$ cells indicates that within the realm of our hypothesis, cells undergoing either trauma or embryogenesis should significantly increase either the amount of factors that they are producing or the spectrum of factors that are produced during these particular states (see Fig. 5). When looking at the listed states we previously outlined relative to one another (Fig. 6), we can see that a cell undergoing normogenesis that reaches a terminal differentiation product 1 (this could be any terminal differentiation state or any intermediate differentiation state) can undergo, in the presence of initiators and tumor promoters, a process we define here as exogenesis or GCD-negative; a process that may also be de-

FIGURE 9 Limit-cycle cell program (signal or gene expression) oscillator model outlining the regulation of multiple programmatic states as a function of cell cycle where k = a normogenic program summary and Kc = a signal summary exceeding the normogenic signal. As k proceeds, its pluripotent amplitude diminishes (A; yellow to violet). The invocation of apoptosis can occur stochastically (B) yielding a temporary outlet or can be released totally from the primary program (C) with reduced probability of collapsing to any normogenic pathway. (Adapted from Wille and Scott 1982, 1984.)

scribed as fractional immortalization leading to a state which is "exogenic." This typically may be the initiated state where significant fidelity is lost but where the lack of fidelity has not reached a state which, in conjunction with subsequent promotion phases, produces a malignant state or exomorphogenic state where the cell now lies outside this terminal differentiation product with respect to its mitotic clock as well as its fidelity. This may be akin to states described by Enoch and Nurse (1990). This loop may be defined as a "permanent" GCD incompetency loop (both apoptotic-incompetent and apoptotic-receptive-incompetent). On the other hand, cells that are in a terminally differentiated state, such as the GCD^+1 state, and where fidelity is still substantially high (whether or not they are exposed to a tumor promoter that is incapable of coupling initiation-induced infidelity or whether they are exposed to an antitumor promoter such as RA, which modifies cell programming in such a manner that the mitotic clocks are profoundly different from the tumor promotion state; this modification cannot couple to an initiation event to produce a malignant phenotype) enter a temporary loop of GCD incompetency (GCD_t^-; blue). These states would typically be metastable with their own defined loops for trauma, antitumor promotion, tumor promotion, and embryogenesis. It is interesting to note at this point that although there appear to be several paradoxical observations relative to this model (e.g., RA is a weak second-stage tumor promoter, and inhibition of the expression of the RA receptor induces a return to apoptosis competency), these paradoxes support our hypothesis that the apoptosis program is only pseudolinked to the mitotic program and that apoptosis is indeed required for the differentiation process, multistate cell rescue, and embryogenesis. Furthermore, these data lend strong evidence for the mechanism of action of compounds such as RA, where RA defines a metastable apoptotic state separate from the driving mitotic clock. In recent experiments by us, our data further indicate that at least in the case of RA and other tumor promoters, relative to times at which cells undergo programmatic setting, these signals can occur extremely fast, and a primary signal involving G proteins, inositol phosphate intermediates, and protein phosphorylation (epigenetic constituents) is coupled to genetic events such as the expression of one or more of the RA receptors which are, in turn, involved in cassette programming and reprogramming with homeobox genes and homeobox gene clusters (Cope 1989; Cope and Wille 1989, 1991).

The Mitotic Clock and Apoptosis

We have provided a significant amount of evidence to indicate that extracellular signals regulate the proliferative and differentiative activities of cells and that these two particular activities exist in profoundly separate states. We would argue at this point, relative to the current school of multistage carcinogenesis, that indeed these separate states are the apoptotic state and the mitotic clock state. These states correspond to the initiation state, where there is a reduction in fidelity that can alternatively be coupled to a necessary coordinating tumor promotion event, which results in a reduction of apoptosis and a concomitant change in the mitotic clock, where the altered clock loop depends on the qualitative and quantitative loss of fidelity, for example, the expression of oncogenes. The role of tumor promoters or antitumor promoters as extracellular signals in regulating the proliferative and differentiative activities of cells has only recently been recognized. The induction of alternate cell states at a restriction point in the G_1 phase of the cell cycle by TPA or RA is outlined in Figure 8. Apoptosis can be mapped to a distinct state in the G_1 phase, here designated GD. GD differs from two other states, GD' and terminal differentiation (TD), which exists in a separate irreversible stratum, whereas GD' is metastable. This is akin to the GCD-competent pathway outlined previously (Fig. 6). Alternatively, a metastable apoptosis-incompetent pathway (RA-induced) is mapped to a separate outlying stratum (GS/C) and may return to the apoptosis-competent pathway via a terminal differentiation state or other intermediate state. However, the neoplastic state (GN) is blocked and remains in orbit (programmatically and mitotically) from the normogenic pathway.

Figure 9A predicts that variable lengths of G_1 rest states in terms of collapse to low amplitudes relative to their oscillations lie within a restricted portion of the G_1 phase structure of the large amplitude oscillation. These variable oscillations predict

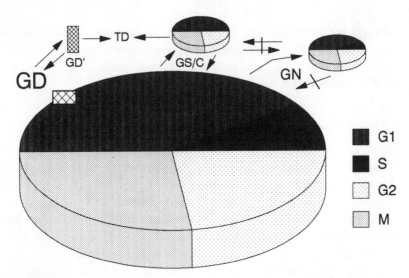

FIGURE 8 Cell-cycle changes in apoptosis: Expression of satellite cycles and their interaccessibility. (See text for description; adapted from Wille and Scott 1982, 1984.)

that any particular state such as we have shown, whether an apoptotic state, particularly one in which cells are apoptosis-competent; tumor promotion where apoptosis is inhibited; antitumor promotion where RA also inhibits apoptosis; and embryogenesis all produce significantly different and unique oscillations that may or may not share elements that define their orbits or potential collapse to final differentiation. These unique structures act in conjunction with gene-cassette programming to alternately define whether or not the state (1) is temporary or permanent, (2) is apoptosis-competent or -incompetent, and (3) is apoptosis-responsive or -unresponsive.

Experimental evaluation of these theoretical possibilities involves testing a number of differing predictions based on perturbation analyses. For example, noncritical perturbations, that is, those given at noncritical phases of the oscillation, drive the system trajectories off the limit-cycle toward the *XY* origin into a region of state space near isochrons that eventually return to the closed oscillatory path without altering the underlying dynamics. The result of such a perturbation or series of perturbations is to recruit a population of asynchronous oscillators (cells) to a narrow range of phases. Then, upon return to permissive conditions, all copies of the oscillation will return to the limiting trajectory with little dispersion in phases (temporary GCD incompetency). On the other hand, a critical perturbation, that is, one given at the critical phase of the oscillation at a critical intensity and duration, drives the oscillatory system toward a singularity in the system, thereby annihilating the oscillation (permanent GCD incompetency).

Finally, according to this interpretation, growth arrest at any state, for example, G_n, is accompanied by the loss of the original high-amplitude oscillation and development of a sub-threshold limit-cycle oscillation (Fig. 9B). Therefore, upon the alterations in signal relative to whether the cell is undergoing embryogenesis, tumor promotion, initiation, or antipromotion, these states can vary in shape and size, and this shape and size relative to the current state of fidelity define temporary versus permanent states and malignant versus normogenic pathways.

SUMMARY AND CONCLUSIONS

The recent paper by Weinstein (1991) was an attempt to indicate the relative potential for carcinogenesis and its linkage to cell proliferation. Many of the points Weinstein (1991) made relate to our previous discussion on the characterization and interrelationships of apoptosis and the malignant phenotype. However, this paper failed to account for the pseudolinkage between the apoptotic state and the mitotic state and the qualitative and quantitative differences in mitotic time structure. Our model, unlike earlier concepts, does not require the reversion of any cell to some primitive dedifferentiated or retrodifferentiated state to explain the appearance of the malignant phenotype. As expressed by Wenner and Tomei (1981), it is possible that the concept of the cancer cell as a stripped-down cell having lost all capabilities of fidelity and attendant biochemical elements directly serving the primary function of terminal differentiation is more correct than the view in which such elements served the primary function of proliferation—in their view, a rather misleading concept. These concepts relate to the separation of apoptosis both qualitatively and quantitatively relative to the modulation of diverse cell-cycle states which in and of themselves are distinct from one another throughout the process of apoptosis, embryogenesis, carcinogenesis, and trauma. This separation not only lends itself well to hypothetical considerations and tests in vivo and in cell culture, but also lends itself remarkably well to intervention for the treatment of malignant disease and/or the prevention of side effects for treatment of malignant disease.

REFERENCES

Alnemri, E.S. and G. Litwack. 1990. Activation of internucleosomal DNA cleavage in human CEM lymphocytes by glucocorticoid and novobiocin: Evidence for a non-Ca^{2+}-requiring mechanism. *J. Biol. Chem.* **265:** 17323.

Altaba, A. Ruiz i and D.A. Melton. 1989. Interaction between peptide growth factors and homeobox genes in the establishment of anterior-posterior polarity in frogs. *Nature* **341:** 33.

Brockes, J. 1991. Retinoic acid may not be a morphogen. *Nature* **350:** 15.

Cope, F.O. 1989. The cytodifferentiating activity of retinoic acid is regulated by IP_4, protein phosphorylation, and G-protein-dependent transduction systems: Key to the universal function of retinoids. In *Bioinformatics: Information transduction and processing systems from cell to whole body* (ed. O. Hatase and J. Wang), p. 141. Elsevier, New York.

Cope, F.O. and J.J. Wille. 1989. Retinoid receptor antisense DNA's inhibit alkaline phosphatase induction and clonogenicity in malignant keratinocytes. *Proc. Natl. Acad. Sci.* **86:** 5590.

———. 1991. Apoptosis appears to be a central path through which the activity of retinoic acid is regulated. *Proc. Am. Assoc. Cancer Res.* **32:** 198.

Enoch, T. and P. Nurse. 1990. Mutation of the fission yeast cell cycle control genes abolishes the dependence of mitosis on DNA replication. *Cell* **60:** 665.

Gillis, P. and J.S. Malter. 1991. The AU-binding factor recognizes the AU-rich elements of cytokine, lymphokine and oncogene mRNAs. *J. Biol. Chem.* **266:** 3172.

Goldstein, S. 1990. Replicative senescence: The human fibroblast coming of age. *Science* **249:** 1129.

Hayflick, L. 1961. The establishment of a line (WISH) of human amnion cell in continuous culture. *Exp. Cell Res.* **23:** 14.

―――. 1965. The limited in vitro lifetime of human diploid cell strains. *Exp. Cell Res.* **37:** 614.

Kyprianou, N., H.F. English, and J.T. Isaacs. 1990. Programmed cell death during regression of PC-82 human prostate cancer following androgen ablation. *Cancer Res.* **50:** 3748.

Ladik, L. 1985. Physical mechanisms of the activation of oncogenes through carcinogens. In *Molecular basis of cancer* (ed. R. Rein), p. 343. Alan Liss, New York.

Lenormand, P., L.L. Muldoon, H. Enslen, K.D. Rodland, and B. Magun. 1990. Two tumor promoters, TPA and thapsigargin act synergistically via distinct signalling pathways to stimulate gene expression. *Cell Growth Differ.* **1:** 627.

Malter, J.S. and Y. Hong. 1991. A redox switch and phosphorylation are involved in the posttranslational up-regulation of the AU-binding factor by phorbol ester and ionophore. *J. Biol. Chem.* **266:** 3167.

Pereira-Smith, O.M. and J.R. Smith. 1988. Genetic analysis of indefinite division in human cells: Identification of four complementation groups. *Proc. Natl. Acad. Sci.* **85:** 6042.

Slawin, K., I.S. Sawczuk, C.A. Olsson, and R. Buttyan. 1990. Chromosomal assignment of the human homologue encoding SGP-2. *Biochem. Biophys. Res. Commun.* **172:** 160.

Sugawara, O., M. Oshmura, M. Koi, L.A. Annab, and J.C. Barret. 1990. Induction of cellular senescence in immortalized cell by human chromosome-1. *Science* **247:** 707.

Takeshita, M., H. Van der Keyl, and A.P. Grollman. 1985. Carcinogen-induced insertion mutations in *E. coli*. In *Molecular basis of cancer* (ed. R. Rein), p. 389. Alan Liss, New York.

Tomei, L.D., P. Kanter, and C.E. Wenner. 1988. Inhibition of radiation-induced apoptosis *in vitro* by tumor promoters. *Biochem. Biophys. Res. Commun.* **155:** 324.

Travali, S., J. Koniecki, S. Petralia, and R. Beserga. 1990. Oncogenes in growth and development. *FASEB J.* **4:** 3209.

Tuck, A.B., S. Wilson, R. Khoka, and A.F. Chambers. 1991. Different patterns of gene expression in *ras*-resistant and *ras*-sensitive cells. *J. Natl. Cancer Inst.* **83:** 485.

Weinstein, I.B. 1991. Mitogenesis is only one factor in carcinogenesis. *Science* **251:** 387.

Wenner, C.E. and L.D. Tomei. 1981. Phenotypic expression of malignant transformation and its relationship to energy metabolism. In *Cell biology monographs* (ed. I.L. Cameron and T.B. Pool), p. 163. Academic Press, New York.

Wille, J.J. and R.E. Scott. 1982. Topography of the predifferentiation GD growth arrest state relative to other growth arrest states in G1 phase of the cell cycle. *J. Cell Physiol.* **112:** 115.

―――. 1984. Cell cycle-dependent integrated control of cell proliferation and differentiation in normal and neoplastic mammalian cells. In *Cell cycle clocks* (ed. L.N. Edmonds), p. 433. Marcel Dekker, New York.

Yuan, J.-Y. and H.R. Horvitz. 1990. The *C. elegans* genes *ced-3* and *ced-4* act autonomously to cause programmed cell death. *Dev. Biol.* **172:** 160.

Apoptosis in the APO-1 System

P.H. Krammer,[1] I. Behrmann,[1] V. Bier,[2] P. Daniel,[1] J. Dhein,[1]
M.H. Falk,[3] G. Garcin,[1] C. Klas,[1] E. Knipping,[1]
K.-M. Lücking-Famira,[1] S. Matzku,[4] A. Oehm,[1] S. Richards,[1]
B.C. Trauth,[1] G.W. Bornkamm,[3] W. Falk,[1] P. Möller,[5]
and K.-M. Debatin[2]

[1]Institute for Immunology and Genetics, German Cancer Research Center,
Heidelberg, Germany
[2]Children's Hospital, University of Heidelberg, Germany
[3]Institute for Clinical Molecular Biology/GSF, Munich, Germany
[4]Institute for Radiobiology and Pathophysiology, German Cancer Research
Center, Heidelberg, Germany
[5]Institute for Pathology, University of Heidelberg, Germany

Cell-surface molecules are crucial in lymphocyte growth control. Such molecules may function as receptors for growth-stimulating cytokines or may be associated with receptors and transmit signals essential for growth regulation. Receptor blockade or removal of the stimulating cytokines may lead to decreased lymphocyte growth (Duke and Cohen 1986). Withdrawal of interleukins slows human lymphocyte growth and finally leads to the characteristic form of "programmed cell death" or apoptosis. Apoptosis is the most common form of eukaryotic cell death and occurs in embryogenesis, metamorphosis, tissue atrophy, and tumor regression. It is also induced by cytotoxic T lymphocytes and natural killer cells, by cytokines like tumor necrosis factor (TNF) and lymphotoxin (LT), and by glucocorticoids. The most characteristic signs of apoptosis are segmentation of the nucleus, condensation of the cytoplasm, membrane blebbing, and DNA fragmentation into multimers of about 180 bp (called a DNA ladder) (see Kerr and Harmon, this volume). To analyze mechanisms of lymphocyte growth control and to interfere with the replication of lymphoid tumor cells, we raised monoclonal antibodies against cell-surface molecules involved in these processes. Monoclonal antibodies were usually tested and selected by virtue of their binding to cell-surface antigens of test cells. Our aim was to define reactive monoclonal antibodies by functional assays, namely by abrogation of growth of malignant test cells in vitro. Monoclonal antibodies were raised against the human B-lymphoblast cell line SKW6.4. One monoclonal antibody, anti-APO-1, showed the strongest functional activity and reacted with an antigen (APO-1) of ~50 kD on a set of activated human lymphocytes, on malignant human lymphocyte lines, and on some patient-derived leukemic cells. Anti-APO-1 was of the IgG3/κ isotype and had a high affinity of K_D = 1.9 x 10^{-10}. Despite many cell fusions undertaken in our laboratory, the hybridoma with anti-APO-1 activity has remained the only one in about 25,000 tested. Nanogram quantities of anti-APO-1 completely blocked proliferation of cells bearing APO-1 in vitro in a manner characteristic of apoptosis (Fig. 1). Cell death was preceded by changes in cell morphology and fragmentation of DNA. This process was distinct from antibody- and complement-dependent cell lysis and was mediated by the antibody alone (Trauth et al. 1989; Krammer 1989; Krammer et al. 1989; Köhler et al. 1990).

Purification of the APO-1 Antigen

It was important to further characterize the APO-1 molecule with the aim of learning more about its function. Therefore, we purified the APO-1 antigen from membranes of SKW6.4 cells. The purified APO-1 antigen was found to be a glycoprotein with apparent M_r of approximately 50,000, with about 8000 of the M_r accounted for by sugars. Purified APO-1 blocked anti-APO-1-induced apoptosis of SKW6.4 cells in vitro, proving its serological identity with the APO-1-membrane antigen. Large quantities of the APO-1 antigen enabled us to obtain a sequence of the APO-1 protein. A computer search revealed that APO-1 was a new cell-surface antigen. Motifs in the APO-1 sequence may provide us with a clue to the as yet elusive physiological function of the antigen.

The APO-1-mediated Signal

Induction of apoptosis was mediated by anti-APO-1 alone and was complement-independent. Nevertheless, the F(ab')$_2$ frag-

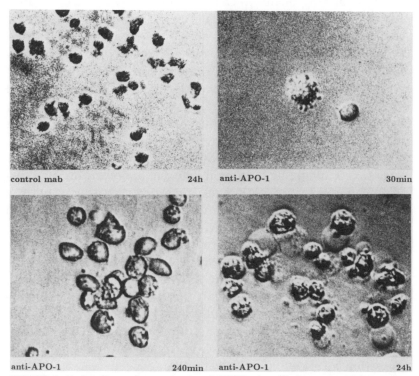

FIGURE 1 Induction of apoptosis of SKW6.4 cells by anti-APO-1. The time of in vitro induction with control monoclonal antibody or anti-APO-1 (1 μg/ml) is indicated.

ment of the IgG3 anti-APO-1 did not induce apoptosis. When cross-linked, however, by F(ab')$_2$ sheep anti-mouse Ig antibodies, apoptosis was observed. To further study the role of the Fc region of anti-APO-1, we isolated antibody class switch variants from the IgG3 anti-APO-1-secreting hybridoma cell line. We obtained anti-APO-1 antibodies of the IgG3, IgG1, IgG2b, IgG2a, and IgA isotypes. These antibodies showed the following effects: (1) a different degree of induction of apoptosis of SKW6.4 cells occurred in the following order: IgG3, IgG1, IgG2a, IgA, IgG2b. (2) Cross-linking of the less effective class switch variant IgG2b anti-APO-1 by Protein A showed the same degree of growth inhibition as IgG3 anti-APO-1. These results suggested that induction of apoptosis was dependent on cross-linking of the APO-1 cell-surface antigen. IgG3 anti-APO-1 bound to the cell surface might have self-aggregating capacity via Fc-Fc interactions or bind to Fc receptors and therefore efficiently cross-link the APO-1 antigen. IgG2b anti-APO-1 might show fewer Fc-Fc interactions, be a less efficient cross-linker, and therefore be less effective in induction of apoptosis. Cross-linking of APO-1 on the cell membrane may be essential for APO-1-mediated signal transduction across the membrane.

We also asked whether internalization of APO-1 and/or anti-APO-1 might be a prerequisite for apoptosis in our system. The following experiments suggested that this is not the case. We chemically coupled anti-APO-1 to silica beads several times larger than cells and incubated SKW6.4 cells with these beads. We found that bead-coupled anti-APO-1 was an efficient inducer of apoptosis in SKW6.4 cells. These results reinforce our assumption that the APO-1 antigen may produce a genuine transmembrane signal, the nature of which remains to be investigated. These results also prompted us to develop systems that might allow us to study the anti-APO-1 apoptosis process in molecular terms. Thus, we looked for cellular systems that might be informative in this respect.

Selection of Cell Variants That Express the APO-1 Antigen but Are Resistant to Anti-APO-1-induced Apoptosis

After screening a large panel of human B- and T-cell lines, we found that expression of the APO-1 antigen is a prerequisite, although not sufficient by itself for anti-APO-1-induced cell death. Thus, we identified several strongly APO-1+ cell lines resistant to anti-APO-1-induced apoptosis. To study this phenomenon further, we selected several cell variants that differed in the sensitivity to anti-APO-1. The B-cell line SKW6.4 (sIgM+, APO-1+, sensitive to 2 ng/ml anti-APO-1) was cultured with increasing amounts of anti-APO-1 for about 1 year. We obtained a stable variant that expressed the APO-1 antigen but was resistant to at least 50 μg/ml anti-APO-1. In addition, the T-cell line CCRF was cloned under limiting dilution conditions. Replica cultures of subclones were screened for susceptibility to anti-APO-1. Two subclones were selected that both expressed the APO-1 antigen but differed in sensitivity to anti-

APO-1 at least by a factor of 1000. It is conceivable that the mechanism of resistance to apoptosis in SKW6.4 and CCRF variant cells is different. In any case, however, this pair of cell lines shows very clearly that two requirements for anti-APO-1-induced apoptosis are important: the cell-surface expression of the APO-1 antigen and an intact apoptosis signal pathway. We presume that these findings may be of great future relevance to the putative use of the apoptosis concept in tumor therapy.

Apoptosis in Human T Lymphocytes

Another informative set of cells with respect to the APO-1-mediated signal of apoptosis are normal human T lymphocytes. Although we have data suggesting that, in contrast to resting B cells, activated B cells also undergo anti-APO-1-mediated apoptosis, in this paper, we focus primarily on T cells. The majority of normal human resting T lymphocytes do not express the APO-1 antigen. After activation, however, both the CD4+ and CD8+ subpopulations of T cells become positive for the APO-1 antigen. Although no significant difference in the amount of APO-1+ T cells and in the epitope density of APO-1 antigens between T cells early (e.g., 1 day) or late (e.g., 6 days) after activation was observed, apoptosis was only induced by anti-APO-1 in the latter cell population. Hence, the susceptibility for induction of apoptosis in activated T lymphocytes is dependent on the stage of differentiation of these cells. A comparison of the set of APO-1+ T cells early or late after activation might help to elucidate the enigma of "death genes" involved in anti-APO-1-mediated apoptosis. In addition, this phenomenon might help to understand in molecular terms the elimination of peripheral T cells at the cessation of an immune response.

Anti-APO-1-mediated Tumor Regression

As discussed above, anti-APO-1 induced apoptosis in various T- and B-cell lines in vitro. This result led us to test the anti-APO-1 efficiency in an experimental tumor system in vivo (Fig. 2). The human B-lymphoma line BJAB was chosen for these in vivo experiments. Xenografts of this line in nu/nu mice were

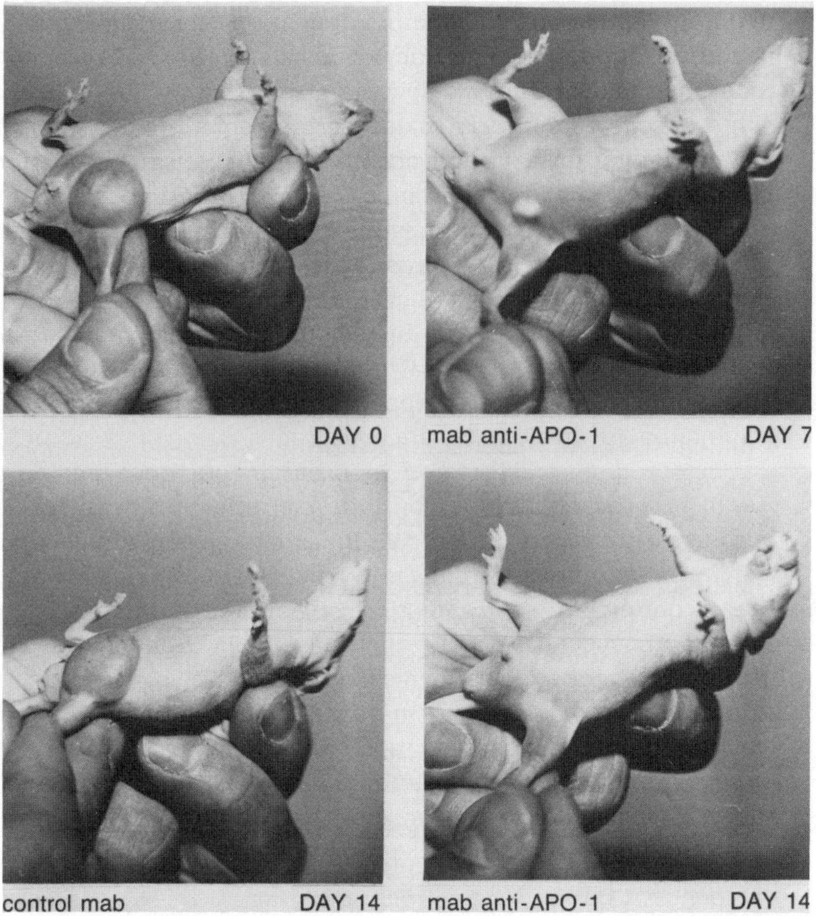

FIGURE 2 Anti-APO-1-mediated tumor regression of BJAB lympho-blastoid tumor xenotransplants in nu/nu mice. The pictures show prototype mice from each group. nu/nu mice with human BJAB lymphoblastoid tumors ~1.5–2.5 cm in diameter (day 0) were i.v. injected with 500 µg isotype matched control monoclonal antibody or anti-APO-1 (IgG3/κ) on day 0. Mice with tumors were photographed 7 and 14 days after monoclonal antibody injection.

previously shown to accumulate radiolabeled monoclonal antibodies only in the outer layer of the tumor, whereas central areas of nodules were virtually inaccessible. Using anti-APO-1 in BJAB-bearing nu/nu mice, we asked three questions: (1) Is anti-APO-1 as effective in vivo as in vitro? (2) Does anti-APO-1 affect the whole tumor despite preferential accumulation in the periphery? (3) Does anti-APO-1-mediated tumor cell death in vivo alter the accessibility barriers of the BJAB tumor? The results were clear-cut. Anti-APO-1 antibodies, like all other antibodies tested, accumulated exclusively in the periphery of nodules even if up to 500 µg of antibody was injected per mouse. Nevertheless, established tumors ~1.5–2.5 cm in diameter regressed in 10/11 nude mice within a few days. Histological thin sections performed before complete tumor regression showed that as in vitro, anti-APO-1 also induced apoptosis in vivo. The action of the antibody, however, did not result in a disturbance of the accessibility barrier. We concluded from these experiments that tumors may be efficiently tackled by monoclonal antibodies, particularly anti-APO-1, despite restriction of accessibility, provided the cytolytic activity of the antibody is high and the residence time of the antibody in the tumor is long enough to "melt down" the tumor nodules from the outside (Trauth et al. 1989). In addition, the outcome of these experiments suggested that anti-APO-1-induced apoptosis is a valid concept worth testing for tumor treatment in a clinical situation, provided putative systemic toxicity of the antibody can be controlled.

One important result should be mentioned at this point. In preliminary experiments, we tested the in vivo therapeutic efficiency of anti-APO-1 on large SKW6.4 tumors. In vitro anti-APO-1-sensitive (S) and -resistant (R) SKW6.4 cells both expressing APO-1 on the cell surface were grown to tumors of about 2 cm in diameter in SCID mice. Anti-APO-1 treatment of these animals resulted in complete tumor regression of the SKW6.4^S tumors only. Animals with SKW6.4^R tumors were killed by the tumor. These results suggested that two requirements for anti-APO-1-mediated tumor regression by induction of apoptosis also exist in vivo: (1) expression of the APO-1 antigen *and* (2) an intact apoptosis signal pathway. As already stated, these results may have far-reaching implications for therapy using rational intervention strategies in the clinic.

Preclinical Applications of Apoptosis in the APO-1 System

The above in vivo experiments prompted us to test APO-1 expression in various tumor systems and to test in vitro induction of apoptosis in malignancies that may be candidates for future anti-APO-1 treatment in the clinic.

Expression of the APO-1 antigen on acute lymphoblastic leukemia cells. In T-acute lymphoblastic leukemia (ALL), APO-1 is expressed constitutively, especially in cases corresponding to stages of very early T-cell differentiation. Cells of the common ALL phenotype representing the malignant precursors of B cells weakly express APO-1 in a minority of cases. However, in these cells, APO-1 expression is induced in vitro by phorbol myristate acetate (PMA) and cytokines such as IL-4. In addition, the constitutive expression of APO-1 on pre-T-ALL cells is modulated by mitogens and cytokines. The APO-1 antigen may therefore be of importance for growth regulation in malignant lymphocytes and may also serve a function in the development of normal precursor cells. In addition, APO-1-positive malignant cells may be a new subgroup of ALL and may be a target for APO-1-directed therapeutic approaches in vitro and in vivo using the anti-APO-1 antibody.

Anti-APO-1 antibody-mediated apoptosis in adult T-cell leukemia. We have described that the APO-1 antigen is expressed on activated T cells and that sensitivity to induction of apoptosis by anti-APO-1 is acquired during long-term culture of activated T cells in the presence of IL-2. Since adult T-cell leukemia (ATL) cells are the transformed counterpart of mature T lymphocytes, we were interested to see whether these cells express the APO-1 antigen and whether they are sensitive to growth inhibition and induction of apoptosis by anti-APO-1. Expression of the antigen and sensitivity to the induction of cell death by anti-APO-1 were studied in human T-cell lines transformed by human leukemia virus type 1 (HTLV-1) and in cultured cells from patients with ATL. APO-1 was strongly expressed on both types of cells, and incubation of the cells with anti-APO-1 resulted in inhibition of proliferation and apoptosis. Induction of apoptosis may therefore be a therapeutic tool in HTLV-1-associated malignant disorders (Debatin et al. 1990).

Expression of the APO-1 phenotype in Burkitt's lymphoma cell lines correlates with a phenotype shift to a lymphoblastoid phe-

notype. We had previously found that APO-1 was also expressed on normal activated B cells (Trauth et al. 1989). Furthermore, a small subset of follicle center B cells residing at a location in which maturation, proliferation, and elimination by apoptosis of B cells takes place had been shown by immunohistochemistry to be APO-1$^+$. Therefore, we tested whether malignant counterparts of such germinal center B cells, Burkitt's lymphoma (BL) cells, expressed APO-1 and were sensitive to anti-APO-1-induced apoptosis. Taking together the evaluation of a large number of tests of BL cells and BL lines phenotypically resembling in vivo BL and cell lines showing a phenotype of Epstein-Barr virus-positive lymphoblastoid cells (LCL), the following results were obtained. BL cells directly isolated from tumor biopsies were APO-1$^-$. BL type cell lines were APO-1$^-$, and LCL type cell lines were APO-1$^+$. Cells of the BL/LCL phenotype showed a heterogeneous APO-1$^+$ pattern. Some but not all cells of the APO-1$^+$ phenotype were sensitive to anti-APO-1-induced apoptosis. The phenotypic shift of BL cell lines may correlate with the one in B-cell activation. Therefore, these cell lines may represent a useful system to study APO-1 expression and function in B cells.

Expression of the APO-1 antigen on glioblastoma cell lines and their susceptibility to apoptosis. To assess the potential usefulness of anti-APO-1 for therapy in other tumor systems, we also tested human glioblastoma cell lines for expression of the APO-1 antigen and susceptibility to anti-APO-1-induced apoptosis. Most cell lines expressed APO-1 at least at a low level. Some cell lines showed growth inhibition and apoptosis if incubated with anti-APO-1. Thus, although APO-1 was expressed on most cell lines tested, only a few responded to anti-APO-1. Subcloning a partially responsive cell line yielded APO-1$^+$, anti-APO-1-sensitive and APO-1$^+$, anti-APO-1-resistant subclones. The data in this cellular system, therefore, stress again that expression of the APO-1 antigen and an intact apoptosis signal pathway are necessary for successful anti-APO-1-mediated apoptosis. Presently, we are investigating which parameters determine the susceptibility of such clones to induction of apoptosis, and whether local anti-APO-1 therapy might be considered in such a disease where survival after relapse is short and no therapeutic possibilities exist.

APO-1 expression in colorectal carcinomas correlates with poor prognosis. All above data on various malignant cells show a common trait. APO-1 expression on the same type of tumor varies. In addition, similar variability is observed as to susceptibility to anti-APO-1-induced apoptosis on APO-1$^+$ malignant cells. Tumors are either sensitive, resistant, or composed of sensitive and resistant cells. This observation also extends to sarcomas and mammary carcinomas not extensively discussed here. Although the physiological function of APO-1 is still unclear, one may speculate that the observed heterogeneity is meaningful for the biology of the tumor and thus also for the clinical course of the malignant disease. These considerations led us to investigate APO-1 expression on colorectal carcinomas and to correlate our findings with the clinical parameters of this malignant disease.

By means of immunohistochemistry, we found that APO-1 is expressed in normal colon epithelium. In a minor fraction of colon adenomas and in 39.6% of colorectal carcinomas, however, APO-1 expression was diminished. In 48.3% of carcinomas, predominantly of the nonmucinous type, APO-1 was completely abrogated. The normal level of APO-1 expression in carcinoma was correlated with the mucinous type ($p<0.0001$). Reduced or lost antigen expression was more frequent in carcinomas localized in the rectum ($p<0.0001$). In a group of 149 patients who had undergone potentially curative surgery for colorectal carcinoma, the physiological level of APO-1 expression was correlated with a shorter survival after relapse ($p = 0.031$) and with an increased risk of tumor-related death ($p = 0.051$) (P. Möller et al., in prep.). This suggested that the APO-1 antigen is important for signals in growth control of normal and malignant cells. Thus, APO-1 may confer growth advantage to malignant cells and determine the grade of malignancy. Furthermore, this first set of clinical data underscores the importance of APO-1 testing and correlation with patient histories in other malignancies. This applies particularly to those in which heterogeneous APO-1 expression is already observed. It would not be surprising if the APO-1 antigen also constituted a valuable prognostic parameter in such diseases.

DISCUSSION AND OUTLOOK

We showed that anti-APO-1 specifically blocked growth and triggered programmed cell death, apoptosis, of a set of activated normal lymphocytes and cells from malignant lymphoid and nonlymphoid lines after binding to the cell-surface protein antigen APO-1. The APO-1 antigen does not seem to be part of the TNF receptor complex, since its representation on the surface of various cells does not correspond to the distribution of TNF receptors; i.e., macrophage cell lines tested so far are APO-1⁻. Nevertheless, it will be important to test whether various apoptosis pathways such as the one triggered by TNF and anti-APO-1 have common features.

Apoptosis is found in all tissues and also in cells from lower organisms. It is conceivable, therefore, that several distinct cell-surface antigens with a different tissue distribution are involved in the induction of apoptosis. Elucidation of the structure of APO-1, its possible connection to the cytoskeleton, and the molecular events following anti-APO-1 binding may resolve some of these issues.

Since APO-1 is expressed on mature activated lymphocytes, additional experiments will be needed to determine whether the antigen might play a role in the down-regulation of the immune response and be involved in selection and elimination of lymphocytes. It has previously been shown that LT, TNF, and killer cells with their effector molecules induce apoptotic cell death. Because anti-APO-1 also induces apoptosis, a number of possibilities might be considered for the physiological role of the APO-1 antigen. APO-1 might be a receptor for cytotoxic molecules or for autocrine growth factors. Alternatively, it could be a molecule essential for vertical or lateral growth signal transduction. Thus, anti-APO-1 might trigger receptors for lytic molecules or block receptors for growth signals. Putative signals given by APO-1 may remain an enigma until the structure of the antigen reveals its secrets. In any case, the elucidation of the APO-1-mediated apoptosis pathway will constitute a challenge for our research and will provide a basis for the development of a rational intervention strategy in various diseases, particularly cancer.

Our data also have clinical relevance. Anti-APO-1 may be useful as a diagnostic tool to define subsets of normal and malignant lymphocytes and other tumor types. In addition, induction of apoptosis may have implications for antitumor therapy. Antibodies have frequently been used as heteroconjugates with toxins or drugs to destroy tumor cells. Our data, however, show that monoclonal antibodies alone can be lethal to target cells, provided these cells express APO-1 and have an intact apoptosis pathway. Anti-APO-1 might, therefore, be considered for ex vivo or in vivo therapy, under conditions where reactivity with vital normal cells can be excluded or tolerated. Thus, in the immediate future, careful toxicity studies in SCID mice reconstituted with a human immune system, in primates, and in patients will be necessary.

It is easily imagined that a successful putative anti-APO-1 therapy might go beyond a therapy of cancer and might involve elimination by apoptosis, e.g., of activated lymphocytes in autoimmune diseases. It should also be considered that apoptosis may be involved in the pathomechanism of the elimination of T-helper lymphocytes in AIDS, a process that is still largely not understood. In this context, we tested the presence of APO-1$^+$ lymphocytes and of anti-APO-1 autoantibodies in AIDS. We found the number of APO-1$^+$ cells increased in HIV$^+$ donors. In addition, in the serum of HIV$^+$ donors, anti-APO-1 autoantibodies were detected. These findings may suggest a role for apoptosis in the depletion of T cells in AIDS and clearly warrant further studies.

Finally, the molecular investigation of cell death induced by anti-APO-1 might lead to a general understanding of apoptosis. In this case, the use of modified or normal physiological ligands to the cell-surface antigen initiating apoptosis or of chemicals interfering with the apoptotic signal might be envisaged.

Taken together, the APO-1 apoptosis system might help to find "death genes" and clarify whether death occurs in steps, is a single-hit event, or can be reversed once its initial signals are triggered. Thus, the investigation of apoptosis shows that essential questions of death are linked and can be as exciting as the essential questions of life.

ACKNOWLEDGMENTS

We thank K. Hexel, G. Hölzl, M. Kaiser, J. Köllner, R. Kühnl, C. Mandl, S. Menges, J. Moyers, and W. Müller for technical assistance; H. Sauter for expert secretarial assistance; D. Hall for organization of the patient follow-up; T. Gernet for help with the biostatistics; and U. Abel, R. Bamford, R. Braun, H.W. Dörr, H. Fischer, C.K Goldmann, E.B. Helm, M. Kiessling, K. Koretz, M. Mercep, H. Näher, A. Peters, D. Petzold, H. Rübsamen-Waigmann, P. Schlag, and T.A. Waldmann for various support and criticisms throughout this study. This study was supported by grants from the tumor center Heidelberg/Mannheim, the Deutsche Krebshilfe (989-91), the Bundesregierung (P1.1-Aids-1075.01, AI02 II-044-88), and the Aids Programm Baden-Württemberg (II-740.1-Aids/41).

REFERENCES

Debatin, K.-M., C.K. Goldmann, R. Bamford, T.A. Waldmann, and P.H. Krammer. 1990. Monoclonal antibody mediated apoptosis in adult T cell leukemia. *Lancet* **335:** 497.

Duke, R.C. and J.J. Cohen. 1986. IL-2 addiction: Withdrawal of growth factor activates a suicide program in dependent T cells. *Lymphokine Res.* **5:** 289.

Köhler, H.-R., J. Dhein, G. Alberti, and P.H. Krammer. 1990. Ultrastructural analysis of apoptosis by the monoclonal antibody anti-APO-1 on a lymphoblastoid B cell line (SKW6.4). *Ultrastruct. Pathol.* **14:** 513.

Krammer, P.H. 1989. Growth control of normal and malignant lymphocytes. *Interdiscip. Sci. Rev.* **14:** 221.

Krammer, P.H., B.C. Trauth, V. Bier, J. Dhein, W. Falk, G. Garcin, C. Klas, W. Müller, A. Oehm, A. Peters, S. Matzku, P. Möller, and K.-M. Debatin. 1989. Apoptosis in monoclonal antibody-induced tumor regression. In *Progress in immunology* (ed. F. Melchers et al.), vol. VII, p. 1104. Springer-Verlag, Berlin.

Trauth, B.C., C. Klas, A.M.J. Peters, S. Matzku, P. Möller, W. Falk, K.-M. Debatin, and P.H. Krammer. 1989. Monoclonal antibody-mediated tumor regression by induction of apoptosis. *Science* **245:** 301.

Apoptosis and Hepatocarcinogenesis

G.M. Ledda-Columbano and A. Columbano
Istituto di Farmacologia e Patologia Biochimica
Università di Cagliari, 09124 Cagliari, Italy

Cell death is a "constitutive" biological phenomenon that may be observed under a variety of conditions, such as development, genetics, teratology, radiology, immunology, and oncology. Despite this fact, the vast majority of papers in the field have described the existence of cell death without many efforts to probe into the mechanism(s) responsible for it. This may be because cell death has long been considered as a "passive" phenomenon or an "accident" caused by the exposure of the cell to an inappropriate environment.

This classic toxicological view of cell death has long predominated despite the finding that a particular mode of cell death defined as apoptosis (Kerr et al. 1972) is intimately related to birth and development, since it is seen during formation of the eggs and sperm, at many vital stages during embryological development, and during morphogenesis and metamorphosis (Glucksmann 1951; Saunders 1966; Wyllie et al. 1980; Beaulaton and Lockshin 1981). Thus, although apoptosis may also occur in response to minor injury, the idea that cell death may be a mechanism for survival rather than destruction has slowly taken hold. This, in turn, has contributed to the recognition that understanding of the mechanism of cell death at a gene level is fundamental to an understanding of the biological process. In the last few years, several studies have focused on genes possibly related to cell death: In the nematode model, the *ced-3* and *ced-4* genes appear to be required for cell death, whereas in deathless mutants, *ced-1*, *ced-2*, and *Nuc-1* are predominantly expressed (Ellis and Horvitz 1986). In mammalian systems, an increased expression of hepatic transglutaminase mRNA has been associated with the occurrence of apoptosis during regression of liver hyperplasia (Fesus et al. 1987). An increased expression of testosterone-repressed prostate message 2 (TRPM-2) has been found in association with apoptosis during involution of rat prostate following castration (Buttyan et al. 1989). Finally, the finding of an increased expression of proto-oncogenes or cell-cycle-related genes such as c-*fos* and c-*myc* during cell death by apoptosis in rat prostate (Buttyan et al. 1988) supports the concept that cell death is an active process that requires activation of specific genes.

Interestingly, the finding of an increased expression of c-*myc* and c-*fos* during the apoptotic process in the prostate suggests that some events occurring during cell death may also take place during cell proliferation. Whether a common signal transduction pathway may modulate or trigger both cell death and cell proliferation is not known, but it would be of great interest to elucidate the signals that activate these genes.

Homeostatic Role of Cell Death in Normal Tissue

A powerful contribution to the recognition that cell death may not necessarily be an accidental phenomenon, but rather an active process, came from the studies by Kerr et al. (1972) and Wyllie et al. (1980). These authors suggested that apoptosis is a normal active phenomenon that plays a fundamental regulatory function in the control of the overall size of the cell population, being complementary but opposite to another fundamental biological process, namely, cell proliferation. A very close relationship between these two processes has been established in several organs and tissues, whereby an excess of cells was created by treatment with mitogenic stimuli; in particular, apoptosis has been suggested to be the mechanism responsible for reducing the excess cell number during the regression of liver parenchymal hyperplasia caused by the mitogens lead nitrate (Columbano et al. 1985) and cyproterone acetate (Bursch et al. 1984), during the regression of renal hyperplasia caused by lead nitrate (Ledda-Columbano et al. 1989), during involution of biliary duct hyperplasia (Bhathal and Gall 1985) produced by ligation of the bile duct or by administration of α-naphthylisothiocyanate (ANIT), and during the regression of hyperplasia of the pancreas after cessation of a diet containing a trypsin inhibitor (Oates et al. 1986). Interestingly, apoptosis does not occur until the mitotic event has taken place, and, after reaching a peak immediately following the maximum increase in the organ size, it rapidly declines and is not detectable once the organ has regained its original size (Columbano et al. 1985; Ledda-Columbano et al. 1989). In addition, it was shown that during the involution of lead-nitrate-induced liver hyperplasia, no elevation in levels of serum glutamate pyruvate transaminase (a marker of cell necrosis) was observed, despite an extensive cell loss (50% of liver DNA was eliminated in a matter of a few days), thus suggesting that leakage of such an enzyme may not occur when hepatocyte death takes the form of apoptosis (Columbano et al. 1985). It would be of interest to investigate whether hepatic necrosis could be discriminated from apoptosis on the basis of the levels of serum transaminases.

Following administration of primary mitogens, the sequence of the main biological events, cell proliferation and cell death, appears to be temporally different from that seen following toxic injury. As shown in Figure 1, in the former case, the pri-

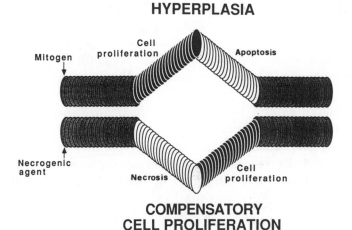

HYPERPLASIA

**COMPENSATORY
CELL PROLIFERATION**

FIGURE 1 Schematic representation of induction of cell proliferation induced by mitogens and necrogenic agents.

mary event is cell proliferation, which results in an excess of cells that are subsequently removed by apoptosis during the regression of the original hyperplasia; in the latter case, the primary event is cell necrosis induced by the hepatotoxicant, and cell proliferation is triggered in order to replace the cells that have been lost. In both cases, a clear relationship between cell proliferation and cell death is evident. Under these conditions, we define as *compensatory cell proliferation* the regenerative process triggered to restore the original number of cells following cell loss; in the same way, we may define as *compensatory cell death* the apoptotic event that eliminates the excess cell number that originates during the initial hyperplasia.

Cell Death and Carcinogenesis

It is now evident that a major reason for the paradoxically slow growth of tumors is the high rate of cell loss (Moore 1987). Contrary to the original perception, Iversen (1967) suggested that although cancer is perceived as a tissue that grows faster than its tissue of origin, it should be more appropriate "to think of a malignant tumor as a tissue that has a slight reduced cell loss." It is therefore conceivable that many tumors survive not simply because of a particularly high rate of cell proliferation, but rather because of a precarious balance of the processes of production and loss. It has been estimated by cell kineticists that approximately 40% of tumor cells are lost, according to the type of tumor studied (Cooper 1973), by exfoliation, migration, or in situ death, the relative contributions of these three modes being dependent on the location of the tumor.

Solid tumors exhibit two forms of cell death: The first is characterized by the presence of individual pycnotic cells scattered among and within otherwise intact parenchymal cells (apoptosis), and the second is characterized by whole regions of dead cells (necrosis). Both forms of cell death may be observed in the same tumor, but the former type may also occur in the absence of the latter (Searle et al. 1973; Moore 1987). During the past two decades, there has been growing awareness of the major importance of apoptosis as a parameter in neoplastic growth (Iversen 1967; Laird 1969; Cooper 1973; Cooper et al. 1975). Apoptotic bodies have been found histologically in several neoplasms (Searle et al. 1975; Kerr and Searle 1972; Cooper et al. 1975; Dempster et al. 1983). Electron microscopy studies have shown that the apoptotic bodies are the same in structure as those derived from non-neoplastic cells and that, although a few of the bodies are taken up by "histiocytes," the majority are rapidly phagocytosed by intact tumoral cells (Kerr et al. 1972). A strict association between apoptotic bodies and mitotic figures was shown in rapidly growing tumors, thus again suggesting that it is the balance between the two processes that determines the rate of enlargement. This finding raises the question whether it is possible to alter the balance between mitosis and apoptosis, so that by increasing the rate of cell loss, one may cause a tumor to regress. Spontaneous occurrence of apoptosis in growing malignant neoplasms appears to be responsible for the regression observed in basal cell carcinomas (Kerr and Searle 1972). In addition, regression of Huggins rat mammary tumors that follows oophorectomy has been shown to be associated with extensive and diffuse apoptotic deletion of tumor cells (Kerr at al. 1972). If it is clear that apoptosis is a fundamental aspect of cancerous cells, its contribution to the growth kinetics of the tumors is difficult to evaluate. Even more important, the role of apoptosis in the growth of neoplasms is not clear. Is it the result of an intrinsic homeostatic mechanism triggered to arrest the growth of the tumor, or is it simply due to the abnormal extracellular environment that pertains in tumor masses? Recently, we have asked whether apoptosis is a late event in the carcinogenic process (possibly restricted to malignant cancer cells) or whether its occurrence could be demonstrated during the various steps of the carcinogenic process, especially the very early steps. Using a model whereby hepatocytes were initiated with a single treatment with the carcinogen 1,2-dimethylhydrazine (1,2-DMH) and promoted to develop into hepatocellular carcinomas by feeding to rats a diet containing 1% orotic acid, we have shown that apoptosis is an early event, since it occurs in preneoplastic foci as early as 10 weeks after orotic acid feeding (Columbano et al. 1984). The presence of apoptotic bodies does not appear to be unique to this model of liver carcinogenesis, because they were also identified in the hepatic foci promoted by other promoting regimens (Table 1). From our studies, it is also evident that their number increased with progression of the process (they were numerous in late nodules, hepatocellular carcinomas, and metastasis) and that their presence was intimately associated with mitotic figures. It is also of interest that the incidence of apoptotic bodies in the hepatocyte nodules was much higher than that observed in the surrounding normal cells (Columbano et al. 1984). The higher susceptibility of preneoplastic hepatocytes to apoptosis is suggestive of a shorter life cycle of these cells. One possible mechanism underlying the observed increase in cell death is that, during carcinogenesis, populations emerge that have difficulty undergoing cell division, and as a result, such cells may die during the division process. Unfortunately, we do not know whether apoptosis during preneoplasia or neoplasia is a random or selective form of cell death. As for normal tissues, it appears that apoptosis occurring during involution of liver hyperplasia (Ledda-Columbano et al. 1991), as well as during pancreas involution (Oates and Morgan 1986), is a random phenomenon in that it eliminates cells that have previously divided as well as cells that did not divide. However, under other conditions, a preferential elimination of cells that did not respond to the mitogenic stimulus has been reported (Bursch et al. 1984).

Another possible explanation is that apoptosis occurs because the nodules as a population represent an overgrowth, and, as such, the same homeostatic mechanism observed in normal tissues may apply for carcinogen-altered cells; however, due to some possible alteration induced by the carcinogen (and/or by the promoting agent), the regulatory mechanism is slightly altered, and the net result is a slow but persistent growth of the preneoplastic lesions. The need to fully elucidate the real mechanism responsible for the occurrence of apoptosis is fundamental. According to the former case, it is possible that cell death created by perturbations in the mitotic apparatus may in turn generate a highly localized compensatory cell proliferation, thus providing an endogenous proliferative stimulus. In this case, apoptosis may favor tumor

TABLE 1 *MODELS OF RAT LIVER CARCINOGENESIS SHOWING APOPTOSIS WITHIN FOCAL OR NODULAR LESIONS*

Initiation	Promotion	Time of investigation	Reference
1,2-DMH	orotic acid	10 weeks to 1 year	Columbano et al. (1984)
1,2-DMH	choline deficiency	6 weeks	Columbano et al. (1984)
1,2-DMH	2-AAF + CCl_4	3 weeks	Columbano et al. (1984)
NNM	phenobarbital	10–28 weeks	Bursch et al. (1984)
DENA	2-AAF + CCl_4	8–28 weeks	Rotstein et al. (1986)
DENA	2-AAF + CCl_4 + phenobarbital	5–16 weeks	Garcea et al. (1989)

(NNM) *N*-Nitrosomorpholine; (2-AAF) 2-acetylaminofluorene.

progression by eliminating "weak" cells and selecting more aggressive cell populations. On the contrary, according to the latter case, the elimination of preneoplastic cells due to a homeostatic mechanism may not imply a compensatory regeneration of adjacent cells, and it may act as a limiting factor for the progression of the process. Is it possible to influence this process? It was suggested that a possible mechanism by which chemicals promote liver carcinogenesis may be to inhibit loss of cells that occurs in preneoplastic lesions (Bursch et al. 1984). Although the removal of the promoter phenobarbital determines an increase in the apoptotic index of hepatic foci, its readministration immediately inhibits the process of apoptosis. Whether the capacity to inhibit cell loss in preneoplastic lesions is common to other promoting agents or whether it is a property unique to phenobarbital (which is also a liver mitogen) remains to be established.

Increasingly, rate of cell loss has also been found in hepatic nodules in the absence of exogenous compound (Rotstein et al. 1986). Interestingly, in this study the increase in growth rate (from 4% at 2 months to 8% at 6 months) was accompanied by an increase in the extent of cell death (from 3% to 7%), thus again suggesting that the balance between the two processes is very critical. It will be of interest to investigate in this model whether induction of apoptosis during the promotion/progression step may influence the development of hepatocellular carcinoma. The study by Rotstein et al. also confirmed previous findings (Bursch et al. 1984; Columbano et al. 1984) on the higher incidence of apoptosis in putative preneoplastic hepatocytes. In fact, it was found that the apoptotic index in surrounding liver was only 0.4% (8 times less than that observed in the nodular hepatocytes). Does the observed higher incidence of apoptotic bodies in preneoplastic lesions truly reflect their shorter half-life, or is it simply the consequence of their slower degradation? Recent studies designed to ask this question revealed that the kinetics of formation and disappearance of preneoplastic apoptotic bodies is similar to that exhibited by apoptotic bodies in liver tissue not exposed to carcinogenic treatment (~ 4 hr) (Bursch et al. 1990).

An aspect that has only recently been considered is the possible role played by apoptosis in the disappearance of preneoplastic lesions. It is known that most of these lesions (up to 95% in some models) disappear after removal of the promoting agent (Enomoto et al. 1982). One mechanism proposed to explain the rapid disappearance of these lesions is the phenomenon defined as "remodeling" or "phenotypic reversion," by which preneoplastic hepatocytes become indistinguishable from the surrounding normal liver cells because they undergo a process of differentiation to adult-like hepatocytes (Tatematsu et al. 1983). This process implies that disappearance of the nodules is not due to cell loss. However, a detailed morphological study during the disappearance of these lesions has not been done. Remodeling may not be the only mechanism involved in this process; concomitant extensive cell loss may also be a contributing factor. A considerable incidence of apoptosis associated with remodeling has been observed in nodular lesions during their disappearance induced by administration of S-adenosyl-L-methionine (Garcea et al. 1989). It is also extremely important to investigate the behavior of products of genes implicated in the apoptotic process. Immunohistochemical analysis associated with the routine histological studies, during different phases of the carcinogenic process, may reveal whether apoptosis is implicated in the disappearance of carcinogen-induced preneoplastic lesions.

Possible Role of Apoptosis during Initiation of Chemical Hepatocarcinogenesis

In the last few years, we have been interested in studying the balance between cell proliferation and cell loss in the various steps of chemical carcinogenesis, in particular, during the initiation and promotion step. As for the first step, it is generally accepted that at least a round of cell replication immediately after or before carcinogen administration is needed for the establishment of an "initiated" cell population (Cayama et al. 1978; Columbano et al. 1981; Ying et al. 1981). The exogenous proliferative stimulus provided in these studies was of compensatory type, such as that elicited by surgical removal of the liver or by a necrogenic dose of CCl_4. As already mentioned (see Fig. 1), when we administer a hepatotoxin such as CCl_4, cell loss occurs in the form of lytic necrosis, and the reduction in cell number triggers a regenerative response of the liver that rapidly restores the original cell number. When this type of proliferative stimulus is applied concomitantly with carcinogen administration, initiation of liver carcinogenesis occurs (Columbano et al. 1987a). Interestingly, when the same dose of carcinogen is coupled with a proliferative stimulus induced by the mitogen lead nitrate, initiation of liver carcinogenesis, monitored as the occurrence of enzyme-altered foci/nodules, does not take place, despite the fact that the extent of cell proliferation is similar to that observed following CCl_4 (Columbano et al. 1987a). The inefficacy of liver hyperplasia to support initiation of hepatocarcinogenesis does not seem to be restricted to lead nitrate, but it appears to be a general phenomenon, since the same pattern of results was obtained using different mitogens, different carcinogens, and different promoting agents (Fig. 2) (Columbano et al. 1987b; Ledda-Columbano et al. 1989).

One may wonder why one type of cell proliferation and not the other supports initiation of carcinogenesis. Is it possible that different types of cell death rather than different proliferative stimuli can explain the different effect on initiation? In particular, is it possible that hepatocytes initiated by the carcinogen during the initial proliferative event are eliminated together with the normal cells by apoptosis during the successive regression phase? (See Fig. 1.) Even if we do not have direct evidence in support of this hypothesis, it is not unrealistic to think that initiated cells may be as susceptible or even more sensitive than normal cells to apoptosis. As mentioned above, we already have evidence that focal and nodular hepatocytes are characterized by a higher incidence of spontaneous cell death; it is possible that one effect produced by the carcinogens is to alter the response of the cell to the apoptotic program. It will be important to study whether initiated cells can be "saved" by conditions that can interfere with the triggering of the apoptotic program.

Possible Role of Apoptosis during Promotion of Chemical Hepatocarcinogenesis

It is generally accepted that the development of initiated cells to foci, nodules, and cancer is based on an as yet unclear ad-

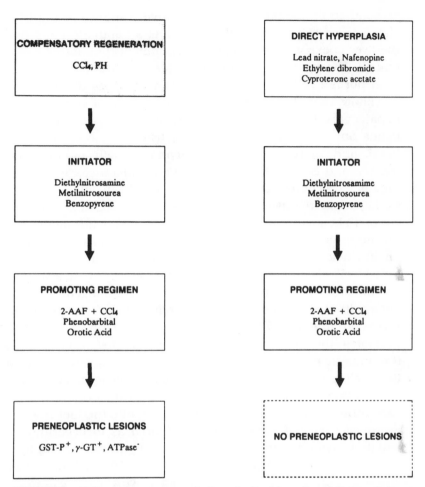

FIGURE 2 Schematic representation indicating the failure of liver mitogens to support initiation of liver carcinogenesis. Rats were treated with nonnecrogenic doses of 3 different carcinogens and exposed to 3 different promoting regimens. Initiation was assayed as the occurrence of foci/nodules showing alteration in the enzyme activity of γ-glutamyltranspeptidase (γ-GT), adenosine triphosphatase (ATPase), and the placental form of glutathione S-transferase (GST-P).

vantage over normal cells. As far as cell death is concerned, for example, it is known that preneoplastic hepatocytes are much more resistant than surrounding normal hepatocytes to severe lytic necrosis induced by several hepatotoxins (Farber et al. 1976). Their resistance to necrosis, coupled with the ability to respond to the consequent regenerative response of the liver, allows them to preferentially expand as a population any time the liver is exposed to a hostile environment (promoting conditions?). The concept that initiated liver cells are more resistant than normal cells may not be true for all types of cell death. We know that the reason for the resistance of preneoplastic hepatocytes to chemically induced cell necrosis greatly depends on their lower capacity to activate xenobiotics on one hand, and on their higher detoxifying capacity on the other (Farber and Sarma 1987). In addition, differences in the uptake of exogenous compounds have also been put forward to explain their relative insensitivity to the necrogenic activity of various chemicals. What about a type of cell death (apoptosis) not necessarily related to toxicity, but rather to expression of an intrinsic homeostatic regulatory mechanism? Will an initiated cell also be resistant to this type of cell death? Our recent studies on promotion of liver carcinogenesis have shown that liver cell necrosis, followed by compensatory regeneration induced by repeated injections with CCl_4, rapidly promotes the growth of initiated cells to foci and nodules (Columbano et al. 1990). On the other hand, repeated waves of liver cell proliferation followed by apoptosis induced by primary mitogens do not promote the appearance of any significant number of preneoplastic lesions, despite the fact that the extent of liver cell proliferation monitored as incorporation of tritiated thymidine into hepatic DNA or by autoradiography is the same in both groups (Columbano et al. 1990). A similar inefficacy of direct hyperplasia associated with apoptosis to develop preneoplastic lesions was also observed when the proliferative stimulus was induced in the presence of promoting agents such as phenobarbital or orotic acid (Fig. 3) (Ledda-Columbano et al. 1991). Why then does development of preneoplastic lesions occur only in the former case (necrosis + compensatory regeneration) and not in the latter (hyperplasia + apoptosis)? One possibility is that initiated cells, unlike normal cells, respond to proliferative stimuli of compensatory type but not to those induced by primary mitogens (Fig. 4). Does this imply that proliferative stimuli of different natures may act through different signal transduction pathways? In this respect, it is interesting to note that liver hyperplasia in-

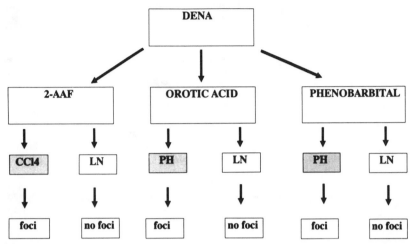

FIGURE 3 Schematic representation indicating that induction of direct hyperplasia induced by mitogens during exposure of rats to three different promoting regimens, unlike compensatory regeneration, does not stimulate the development of enzyme-altered foci. After initiation with diethylnitrosamine (DENA), rats were exposed to 2-AAF, orotic acid, or phenobarbital. Cell proliferation was induced by CCl_4 or partial hepatectomy (PH) (compensatory regeneration) or lead nitrate (LN) (direct hyperplasia) during treatment with the promoting agents.

duced by mitogens may occur in the absence of any increase in the expression of proto-oncogenes such as c-*fos* and c-*myc* (Coni et al. 1990), indicating that, unlike conditions characterized by compensatory regeneration, their enhanced expression may not represent a necessary prerequisite for the entry of the cell into the cycle. Another possibility is that initiated cells respond as much or even more than normal cells to the proliferative stimuli elicited by primary mitogens, but they die during the subsequent apoptotic wave occurring during the regression of the hyperplastic liver (Fig. 4). If this is the case, it is obvious that initiated cells will never reach a focal or a nodular stage. This possibility assumes that initiated cells, despite the carcinogen-induced alterations, are still subject to the same mechanism that so strictly regulates the size of the normal tissues. We believe that it is now fundamental to concentrate the efforts to test this possibility. So far, the strategy underlying chemotherapy has been the inhibition of DNA syn-

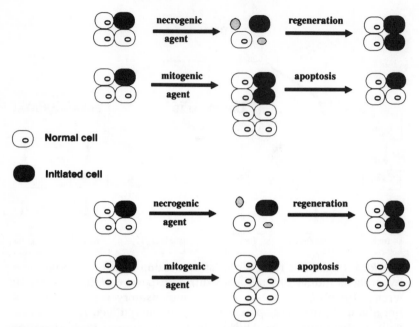

FIGURE 4 Schematic representation indicating the possible mechanism responsible for the difference in the promoting efficacy of compensatory cell proliferation as opposed to direct hyperplasia.

thesis and, consequently, of the proliferation of preneoplastic or neoplastic cells. This approach, however, is limited by several factors: Only a certain fraction of tumoral cells are dividing at a certain time; tumoral cells rapidly develop resistance to anticancer drugs; anticancer drugs are very toxic for the normal cells; etc. On the contrary, if initiated cells are susceptible to apoptosis (we know that the apoptotic index is higher in preneoplastic lesions when compared to normal surrounding cells), is it possible to paradoxically eliminate initiated cells by inducing cell division by primary mitogens in order to delete them by apoptosis during the successive phase of regression? In this respect, it should be stressed that a rapid disappearance of preneoplastic lesions has been observed in rat liver following exposure of the animals to regimens containing mitogenic compounds (De Angelo et al. 1984; Perera and Shinozuka 1984; Staubli et al. 1984), which suggests that the accelerated regression of the lesions could have been caused by an increased cell deletion (apoptosis) of preneoplastic cells.

Although we do not have direct evidence for the proposed role of apoptosis during liver carcinogenesis, we now have the tools to probe into this aspect. By an accurate morphological analysis, we can now start to investigate the response of preneoplastic lesions to chemicals that induce apoptosis, focusing more on the process rather than on the behavior of the chemical; possibly we will be able to determine to what extent focal and nodular cells or normal surrounding cells are sensitive to this type of cell death and how the fate of these lesions is modified by apoptosis. In addition, by using the appropriate models, we may also study the apoptotic process within nodular hepatocytes at a molecular level by isolating the lesions from the normal tissue and examining the expression of genes such as TGase or *TRPM-2* during the several steps of hepatocarcinogenesis and comparing it with that of the normal surrounding hepatocytes. Techniques such as in situ hybridization may even allow us to detect early events in the apoptotic cycle at a cellular level in each step of the carcinogenic process. By studying the various steps of the carcinogenic process, we may also try to ascertain whether transition from preneoplasia to malignancy is the result of a failure of a system for self-destruction.

These are only a few of the questions that need to be resolved, but as stated by Lockshin and Bowen (1981), "It is gratifying to see how similar are the thoughts from the different fields." Undoubtedly, the strikingly growing interest in the mechanisms related to apoptosis registered in the last couple of years and the collaborations already started between groups involved in different fields will greatly improve our knowledge of this biological phenomenon.

REFERENCES

Beaulaton, J. and R.A. Lockshin. 1982. The relationship of programmed cell death to development and reproduction: Comparative studies and an attempt at classification. *Int. Rev. Cytol.* **79:** 215.

Bhathal, P.S., and J.A.M. Gall. 1985. Deletion of hyperplastic biliary epithelial cells by apoptosis following removal of the proliferative stimulus. *Liver* **5**: 311.

Bursch, W., S. Paffe, B. Putz, G. Barthel, and R. Schulte-Hermann. 1990. Determination of the length of the histological stages of apoptosis in normal liver and in altered hepatic foci of rats. *Carcinogenesis* **5**: 847.

Bursch, W., B. Lauer, I. Timmermann-Trosiener, G. Barthel, J. Schuppler, and R. Schulte-Hermann. 1984. Controlled death (apoptosis) of normal and putative preneoplastic cells in rat liver following withdrawal of tumor promoters. *Carcinogenesis* **5**: 453.

Buttyan, R., Z. Zakeri, R.A. Lockshin, and D. Wolgemuth. 1988. Cascade induction of c-*fos*, c-*myc*, and heat shock 70 k transcripts during regression of the rat ventral prostate gland. *Mol. Endocrinol.* **2**: 650.

Buttyan, R., C.A. Olssen, J. Pinter, C. Chang, M. Bandyk, P.-Y. Ng, and I.S. Sawczuk. 1989. Induction of TRPM-2 gene in cells undergoing programmed cell death. *Mol. Cell. Biol.* **9**: 3473.

Cayama, E., H. Tsusa, D.S.R. Sarma, and E. Farber. 1978. Initiation of chemical carcinogenesis requires cell proliferation. *Nature* **275**: 60.

Columbano, A., S. Rajalakshmi, and D.S.R. Sarma. 1981. Requirement of cell proliferation for the initiation of liver carcinogenesis as assayed by three different procedures. *Cancer Res.* **41**: 2079.

Columbano, A., G.M. Ledda-Columbano, P. Coni, and P. Pani. 1987a. Failure of mitogen-induced cell proliferation to achieve initiation of rat liver carcinogenesis. *Carcinogenesis* **8**: 345.

Columbano, A., G.M. Ledda-Columbano, G. Lee, S. Rajalakshmi, and D.S.R. Sarma. 1987b. Inability of mitogen-induced liver hyperplasia to support the induction of enzyme-altered islands induced by liver carcinogens. *Cancer Res.* **47**: 5557.

Columbano, A., G.M. Ledda-Columbano, P.M. Rao, S. Rajalakshmi, and D.S.R. Sarma. 1984. Occurrence of cell death (apoptosis) in preneoplastic and neoplastic liver cells: A sequential study. *Am. J. Pathol.* **116**: 441.

Columbano, A., G.M. Ledda-Columbano, M.G. Ennas, M. Curto, A. Chelo, and P. Pani. 1990. Cell proliferation and promotion of rat liver carcinogenesis: Different effect of hepatic regeneration and mitogen induced hyperplasia on the development of enzyme-altered foci. *Carcinogenesis* **5**: 776.

Columbano, A., G.M. Ledda-Columbano, P. Coni, G. Faa, C. Liguori, G. Santacruz, and P. Pani. 1985. Occurrence of cell death (apoptosis) during involution of liver hyperplasia. *Lab. Invest.* **52**: 670.

Coni, P., G. Pichiri-Coni, G.M. Ledda-Columbano, P.M. Rao, S. Rajalakshmi, D.S.R. Sarma, and A. Columbano. 1990. Liver hyperplasia is not necessarily associated with increased expression of c-*fos* and c-*myc* mRNA. *Carcinogenesis* **5**: 835.

Cooper, E.H. 1973. The biology of cell death in tumours. *Cell Tissue Kinet.* **6:** 87.

Cooper, E.H., A.J. Bedford, and T.E. Kenny. 1975. Cell death in normal and malignant tissues. *Adv. Cancer Res.* **21:** 59.

De Angelo, A.B., C.T. Garret, and A.E. Queral. 1984. Inhibition of phenobarbital and dietary choline deficiency promoted preneoplastic lesions in rat liver by environmental contaminant di(2-ethylhexyl-)phtalate. *Cancer Lett.* **13:** 323.

Dempster, A.G., W.R. Lee, S. Bahnasawi, and T. Downie. 1983. Cell necrosis and endocytosis (apoptosis) in an embryonal rhabdomyosarcoma of the orbit. *Graefe's Arch. Clin. Exp. Opthalmol.* **221:** 89.

Ellis, H.M. and H.R. Horvitz. 1986. Genetic control of programmed cell death in the nematode *C. elegans. Cell* **44:** 817.

Enomoto, K. and E. Farber. 1982. Kinetics of phenotypic maturation of remodeling of hyperplastic nodules during liver carcinogenesis. *Cancer Res.* **42:** 2330.

Farber, E. and D.S.R. Sarma. 1987. Biology of disease. Hepatocarcinogenesis: A dynamic cellular perspective. *Lab. Invest.* **56:** 4.

Farber, E., S. Parker, and M. Gruenstein. 1976. The resistance of putative premalignant liver cell populations, hyperplastic nodules, to the acute cytotoxic effects of some hepatocarcinogens. *Cancer Res.* **36:** 3879.

Fesus, L., V. Thomazy, and A. Falus. 1987. Induction and activation of tissue transglutaminase during programmed cell death. *FEBS Lett.* **224:** 104.

Garcea, R., L. Daino, R. Pascale, M.M. Simile, M. Puddu, S. Frassetto, P. Cozzolino, M.A. Seddaiu, L. Gaspa, and F. Feo. 1989. Inhibition of promotion and persistent nodule growth by S-adenosyl-L-methionine in rat liver carcinogenesis: Role of remodeling and apoptosis. *Cancer Res.* **49:** 1850.

Glucksmann, A. 1951. Cell deaths in normal vertebrate ontogeny. *Biol. Rev.* **26:** 59.

Iversen, O.H. 1967. Kinetics of cell proliferation and cell loss in human carcinomas. *Eur. J. Cancer* **3:** 389.

Kerr, J.F.R. and J. Searle. 1972. A suggested explanation for the paradoxically slow growth rate of basal cell carcinomas that contain numerous mitotic figures. *J. Pathol.* **107:** 41.

Kerr, J.F.R., A.H. Wyllie, and A.R. Currie. 1972. Apoptosis: A basic biological phenomenon with wide ranging implications in tissue kinetics. *Br. J. Cancer* **26:** 239.

Laird, A.K. 1969. Dynamics of growth in tumors and in normal organisms. *Natl. Cancer Inst. Monogr.* **30:** 15.

Ledda-Columbano, G.M., A. Columbano, P. Coni, G. Faa, and P. Pani. 1989. Cell deletion by apoptosis during regression of renal hyperplasia. *Am. J. Pathol.* **135:** 657.

Ledda-Columbano, G.M., P. Coni, M. Curto, P. Pani, D.S.R. Sarma, and A. Columbano. 1991. Cell death and cell proliferation in experimental hepatocarcinogenesis. In *Modulating factors in multi-stage chemical carcinogenesis* (ed. A. Columbano et al.). Plenum Press, New York. (In press.)

Ledda-Columbano, G.M., A. Columbano, M. Curto, M.G. Ennas, P. Coni, D.S.R. Sarma, and P. Pani. 1989. Further evidence that mitogen-induced cell proliferation does not support the formation of enzyme-altered islands in rat liver by carcinogens. *Carcinogenesis* **10:** 847.

Lockshin, R.A. and I.D. Bowen. 1981. Preface. In *Cell death in biology and pathology* (ed. I.D. Bowen and R.A. Lockshin), p. xvii. Chapman and Hall, New York.

Moore, J.V. 1987. Death of cells and necrosis of tumours. In *Perspectives on mammalian cell death* (ed. C.S. Potten), p. 295. Oxford Science Publications, England.

Oates, P.S. and R.G.H. Morgan. 1986. Random or selective cell death during pancreatic involution following withdrawal of raw soya flour feeding in the rat. *Pathology* **18:** 234.

Oates, P.S., R.G.H. Morgan, and A.M. Light. 1986. Cell death (apoptosis) during pancreatic involution after raw soya flour feeding in the rat. *Am. J. Physiol.* **250:** G9.

Perera, M.I.R. and H. Shinozuka. 1984. Accelerated regression of carcinogen-induced preneoplastic hepatocyte foci by peroxisome proliferators BR931, 4-chloro-6-(2,3-xylidino)-2-pyrimidinylthio (N-B-hydroxyethyl)acetamide and di(2-ethylhexyl)phtalate. *Carcinogenesis* **5:** 1193.

Rotstein, J., D.S.R. Sarma, and E. Farber. 1986. Sequential alterations in growth control and cell dynamics of rat hepatocytes in early precancerous steps in hepatocarcinogenesis. *Cancer Res.* **46:** 2377.

Saunders, J.W. 1966. Death in embryonic systems. *Science* **154:** 604.

Searle, J., D.J. Collins, B. Harmon, and J.F.R. Kerr. 1973. The spontaneous occurrence of apoptosis in squamous carcinomas of the uterine cervix. *Pathology* **5:** 163.

Searle, J., T.A. Lawson, P.J. Abbott, B. Harmon, and J.F.R. Kerr. 1975. An electron-microscope study of the mode of cell death induced by cancer chemotherapeutic agents in population of proliferating normal and neoplastic cells. *J. Pathol.* **116:** 129.

Staubli, W., P. Bentley, F. Bieri, E. Frohlich, and F. Waechter. 1984. Inhibitory effect of nafenopin upon the development of diethylnitrosamine induced enzyme-altered foci within rat liver. *Carcinogenesis* **5:** 41.

Tatematsu, M., Y. Nagamine, and E. Farber. 1983. Redifferentiation as a basis for remodeling of carcinogen-induced hepatocyte nodules to normal appearing liver. *Cancer Res.* **43:** 5049.

Wyllie, A.H., J.F.R. Kerr, and A.R. Currie. 1980. Cell death: The significance of apoptosis. *Int. Rev. Cytol.* **68:** 251.

Ying, T.S., D.S.R. Sarma, and E. Farber. 1981. Role of acute hepatic necrosis in the induction of early steps in liver carcinogenesis by diethylnitrosamine. *Cancer Res.* **41:** 2096.

Cell Death in Cancer Chemotherapy: The Case of Adriamycin

T.R. Tritton

Department of Pharmacology and Vermont Regional Cancer Center
University of Vermont College of Medicine
Burlington, Vermont 05405

INTRODUCTION TO THE PROBLEM

The objective of cancer treatment is to kill cancer cells without killing the host. Virtually all successful approaches toward the management of neoplastic diseases rely on techniques that produce cell death in the offending tumor cells. In surgery, the physician kills the cells by physically cutting them away. In radiotherapy, the toll is exacted by killing the exposed cells with ionizing radiation. The third major type of treatment, chemotherapy, brings about cell death by the direct application of chemical poisons, albeit poisons that have some degree of selectivity for tumors. Viewed in this way, the recognition of cell death as a central goal provides a unifying theme for the treatment of cancer. Even newer approaches, particularly immunotherapy, attempt to provoke cell death by taking advantage of intrinsic or natural toxic mechanisms; these could involve apoptosis because, as Kerr and Harmon discuss elsewhere in this volume, gene-directed cell death occurs spontaneously in all tumors, and enhancement of this process could have profound implications for treatment.

In this paper, we focus on chemotherapy and one drug in particular, adriamycin. This agent is one of the most useful single drugs employed to treat human cancer and, because of this central importance, there has been considerable effort devoted to understanding its mechanism of action. For our purposes here, mechanism of action is taken to mean mechanism of causing cell death, and only recently has this come to encompass the possibility of specific gene expression in the process (Posada et al. 1989a; Bhushan et al. 1991). The use of the term cell death is rather new to the chemotherapy field; historically, the descriptive term used was cytotoxicity, and virtually all the important anticancer agents were discovered because they are cytotoxic. Mainly because of a lack of systematic evidence, it is not yet clear whether the pathway initiated by adriamycin (or most of the other antineoplastic drugs) conforms to the classic passage through apoptosis, although one expects directed research on this question will quickly provide answers. Not wishing to offend those who are seeking rigorous definitions of key terms, we nonetheless use the terms cytotoxicity and cell death interchangeably and take loss of reproductive capacity as the measure of death. Thus, operationally a cell is killed by adriamycin if it can no longer divide to form a colony of progeny cells. Other workers in the cancer chemotherapy field have employed diverse measures of cytotoxicity, including dye exclusion, lactate dehydrogenase or chromium release, and decrease in cell number. Elsewhere in this volume, many other criteria for cell death are examined and discussed, but the simple definition of loss of clonogenic capacity serves us well in our quest to define adriamycin's mechanism of action and also keeps tradition with the histological terminology used by most cancer chemotherapists.

There are many theories to explain adriamycin's action (Arcamone 1981; Gianni et al. 1983; Tritton and Hickman 1985; Tritton 1991) and, as we shall see, not all are mutually exclusive. Nonetheless, the baffling array of biological actions wrought by this drug has made it difficult to pinpoint the actual critical lesion that leads to cell death. It seems evident, however, that before one can know *how* the drug works, one must know *where* it works. Even this problem has seemed intractable, however, because adriamycin acts at so many sites that it has proven difficult to separate the crucial from the merely accidental. One approach has been to attempt to separate all intracellular sites from those residing on the outside, or cell membrane. A way to accomplish this goal has been to prepare immobilized or polymeric forms of the drug that cannot enter cells, thereby providing reagents that can effectively distinguish between action at the cell surface and action in the interior. We have recently reviewed this approach (Tritton 1991) and will not repeat the details here. Suffice it to say that although it can be convincingly established that immobilized adriamycin can kill cells from the surface, this does not prove that native adriamycin acts similarly. Therefore, recent work in our laboratory has sought another way to learn which target locus is essential for adriamycin-induced cell death. We have found that temperature is a very useful variable in this regard, and this subject will be the basis for the next section.

Temperature Dependence of Cell Death Caused by Adriamycin

Long-standing tradition in the anthracycline field held that adriamycin is not taken up by cells at low temperature, generally defined as 0°C. This suggests an experiment: Expose cells to drug at 0°C and if the drug has to get inside to cause cell death, it should be ineffective at the low temperature. Conversely, if drug action occurs at the cell surface, cytotoxicity should still occur even at 0°C. Although this experiment seems both logical and simple in its design, it is based on an incorrect assumption; nonetheless, it still yielded useful and interesting results (Lane et al. 1987). Figure 1 shows the incorrect assumption that no drug will be taken inside the cells at 0°C. Although uptake does decline systematically with lowering the temperature, there is still measurable accumulation in the cells even at 0°C. Thus, the original basis for this approach—that drug would interact only with the cell surface at low temperature—was faulty.

Even though adriamycin does gain access to the cell interior at 0°C, when we measured the temperature dependence of cell death, we were surprised to find the temperature profile shown in Figure 1. At all temperatures below about 20°C, there is no cytotoxic response to the presence of the drug. In fact, no matter how high the adriamycin concentration or how long the exposure, as long as the temperature is below the critical 20°C, there is no untoward cellular response to the presence of this otherwise noxious agent (we note that the APO-1 antigen discussed by Dr. Krammer et al. [this volume] shows a similar temperature response). By itself this result

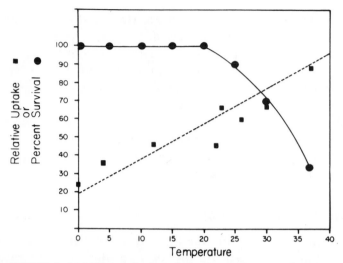

FIGURE 1 L1210 survival and drug uptake as a function of temperature. Survival was determined by cloning in soft agar and uptake was determined by fluorescence determination of extracted cells; in each case, the exposure was to 10^{-7} adriamycin for 2 hr. (Reprinted, with permission, from Lane et al. 1987.)

does not prove anything, but if one assumes that the critical target for adriamycin is inside the cell, there should be a direct relationship between uptake and cell death (i.e., the more uptake, the more likely the cell will die). Clearly, this relationship does not hold, so from this result, one is inclined to favor the extracellular (plasma membrane) target for drug action, since this location would not be expected to depend on intracellular uptake.

If uptake does not govern cytotoxicity, something else must, since the drug is surely capable of causing cell death. Table 1 lists several variables we have examined in an attempt to rationalize the unusual temperature dependence of adriamycin's action (summarized from Lane et al. 1987; Vichi et al. 1989). One obvious possibility is metabolism: By this explanation, the drug would be converted to a toxic species by cellular processes at 37°C, but not at low temperature. This is not the case, however, since the only biotransformation of the native drug in these L1210 (and most other cultured) cells is cleavage to a sugar and aglycone moiety. This reaction occurs only to

TABLE 1   *VARIABLES EXAMINED TO EXPLAIN THE TEMPERATURE DEPENDENCE OF ADRIAMYCIN'S ABILITY TO CAUSE CELL DEATH*

Variable	Correlation with cytotoxicity temperature profile
Uptake	no
Metabolic conversion	no
Subcellular distribution	no
Redox reactions	no
Membrane fluidity	yes
DNA damage	yes

the extent of a few percent, produces only inactive species, and shows little change with temperature. Thus, altered metabolic conversion is an unlikely explanation for the low-temperature inactivity of adriamycin.

A second possible explanation for the temperature dependence is altered subcellular distribution. In this scenario, shifting the cells to low temperature would prevent the drug from gaining access to its critical target. This too is unlikely, however, because both fluorescence microscopy and subcellular fractionation reveal no important differences in the distribution of drug among different compartments when comparing high and low temperature. Moreover, a cytotoxic insult begun at 37°C can be immediately halted at any time by simply shifting to low temperature (P. Vichi and T.R. Tritton, in prep.); it seems implausible (although not impossible) that the equilibrium distribution of drug inside the cell would be so rapidly reversible.

The third possibility listed in Table 1 to explain the temperature dependence of adriamycin's ability to cause cell death is redox reactions. The anthracyclines are quinone-containing molecules and, as such, can undergo one- to two-electron reductions to produce reactive oxygen or alkylating species (Bachur et al. 1979). If these reactions are important to cytotoxicity, and if they do not occur at low temperature, this could explain the temperature profile of drug action. We have previously shown that adriamycin is a perfectly capable cytotoxin in the absence of O_2, suggesting that the aforementioned redox reactions are not vital to the cell death process

(Kennedy et al. 1983). In addition, the NADH- and NADPH-dependent enzymes that can catalyze the reactions are perfectly functional at low temperature (Lane et al. 1987) and show no unusual Arrhenius behavior near the critical 20°C temperature. Consequently, we think that altered redox capability is not the explanation to low-temperature adriamycin inactivity.

One cellular property we find that does show a temperature dependence reminiscent of the cell death profile in Figure 1 is the fluidity of the plasma membrane. The temperature dependence of adriamycin intrinsic fluorescence polarization when bound to the plasma membrane of L1210 cells clearly shows a discontinuity at about 20°C (Lane et al. 1987). Such behavior is generally taken as presumptive evidence for a membrane phase change (solid↔fluid transition) at the indicated temperature. Thus, the binding site(s) for drug on the cell surface does exhibit a structural change at the same temperature that the drug becomes incapable of causing cell death. This could be a coincidence, but we think it equally likely that the structure of the plasma membrane is involved in regulating the response to drug; this idea will be more fully developed later.

There is considerable sentiment among cancer chemotherapies that DNA is a major target for adriamycin. The drug does bind to DNA by intercalation and does cause DNA damage, so we decided to examine the temperature dependence of DNA lesions to see if this property resembled the temperature dependence of cell death (Vichi et al. 1989). We note that, as of this writing (September 1990), there are no reports of adriamycin causing the formation of the nucleosome fragments known as DNA ladders that very often accompany programmed cell death. Such a response may or may not occur with adriamycin, but it has not been systematically investigated for this drug (the intensity of interest in apoptosis may invalidate this assertion even before publication of this volume). Most laboratories measure adriamycin-induced DNA damage by alkaline elution. As shown in Table 2, the temperature dependence of formation for both single-strand breaks and DNA-protein cross-links is identical to the temperature dependence of cell death; i.e., above 20°C, there is DNA damage and it is accompanied by cell death, whereas below 20°C, there is neither

TABLE 2 *DNA DAMAGE INDUCED BY ADRIAMYCIN TO WHOLE CELLS AND ISO-LATED NUCLEI AT HIGH AND LOW TEMPERATURE*

	Rad equivalents DNA damage	
	37 °C	0 °C
Whole cells	420	0
Isolated nuclei	380	89

L1210 cells were exposed to 6 µM adriamycin for 2 hr, and DNA damage was assessed by alkaline elution. The results shown represent single-strand breaks; similar results are obtained if one measures DNA-protein cross-links (Vichi et al. 1989).

DNA damage nor cell death. Thus, these results strongly suggest a functional linkage between adriamycin's ability to derange DNA and its ability to kill the cell. Such a conclusion is surprising to us for two reasons. First, the evidence discussed here so far lies more in favor of the plasma membrane than of intracellular DNA as the primary target for adriamycin, and second, other investigators have been unable to show a quantitative relationship between DNA damage and cell death for adriamycin (Ross et al. 1979; Zwelling et al. 1982). Nonetheless, these results suggest that there is a relationship, and this fact must therefore be considered as we attempt to define how adriamycin causes the cytotoxic response.

DNA is not damaged by adriamycin itself, but by the aberrant action of the enzyme topoisomerase II (Tewey et al. 1984). This protein participates in DNA replication and transcription by relieving topological constraints in the structure of the double helix. It does so by covalently attaching to the nucleic acid, opening the strands to allow passage, and then resealing the duplex. Malfunction of this enzyme yields both DNA breaks and stabilized DNA-topoisomerase complexes. Thus, another potential explanation for adriamycin's inability to kill cells at low temperature would be that topoisomerase II becomes inactive when cooled. We tested this possibility (Vichi et al. 1989) by measuring the ability of topoisomerase to act on two different substrates: knotted P4 DNA and superhelical pBR322 DNA. The results show that the enzyme is functional at low temperature (albeit with a reduced catalytic rate) and that adriamycin can still disrupt its ability to process DNA correctly. Consequently, altered topoisomerase II activity cannot explain the temperature dependence of DNA damage and cell death wrought by adriamycin.

We now have an apparent paradox. Adriamycin cannot kill cells at low temperature nor provoke DNA damage, but the enzyme that causes this damage remains perfectly functional, even at 0°C. A way to rationalize this dilemma would be to postulate the existence of another cellular factor that regulates topoisomerase II activity. To assess this possibility, we isolated cell nuclei, discarding all cytoplasmic and membrane components that might interact with topoisomerase II. Table 2 shows that adriamycin does cause DNA damage in these isolated nuclei at 37°C and, unlike with whole cells, also at 0°C. The DNA damage in isolated nuclei at 0°C is both time and dose dependent, as one would expect. Thus, purifying the nuclei does appear to eliminate a factor that causes the DNA to become unresponsive to damage by adriamycin in intact cells. The identity of this factor is not known, but the possibility that it is protein kinase C (PKC) is discussed in the next section.

We realized that the temperature properties of the cellular response to adriamycin offered a means to place drug on the inside of cells but not the outside. This can be accomplished by incubating the cells at low temperature with adriamycin, followed by washing away the external drug. In principle, the cells can be loaded with any desired concentration, since the drug will be inactive as long as the temperature is maintained below 20°C. With the cell full of drug at 0°C, a temperature shift to 37°C then allows us to assess whether or not there is cell death with drug located on the inside. A complication exists: The drug will equilibrate to a new distribution when the temperature is raised, and thus some efflux is expected. To account for this, we must measure the adriamycin concentration both inside and outside the cell under all conditions. Table 3 provides a condensed summary of this exercise (P. Vichi and T.R. Tritton, in prep.): The results show that when cells are loaded with an intracellular concentration of 10 μM drug at 0°C and washed free of external drug, there is no ensuing cell death when the temperature is shifted to 37°C, even though the concentration achieved inside the cells is four times higher than necessary to cause cytotoxicity when given directly at

TABLE 3 CYTOTOXICITY CAUSED BY ADRIAMYCIN WHEN CELLS ARE LOADED WITH DRUG AT 0 °C, WASHED FREE OF EXTERNAL DRUG, AND CLONED AT 37 °C

Condition of drug exposure	Cytotoxicity at 37 °C ?
A. 37°C, 2.5 μM intracellular	yes
B. 0°C, 10 μM intracellular	no
C. 0°C, 40 μM intracellular	yes
D. 0°C, 40 μM intracellular +/external DNA	no

37°C (condition B). Under these conditions, drug efflux results in a 3-nM adriamycin concentration in the medium. If the intracellular loading is raised to 40 μM at 0°C and then washed and shifted to 37°C, now there are cell death sequelae (condition C). Thus, it appears that drug *inside* the cells is not by itself sufficient to initiate the cytotoxic cascade, but that extracellular drug is also required, and this concentration must exceed a certain threshold.

We reasoned that if it is true that drug originally loaded at 0°C must leave the cell and perturb the surface in order to produce cell death at 37°C, then removal of the effluxed component should prevent cytotoxicity. We tested this idea by providing a high-capacity binding site for adriamycin in the extracellular milieu; this trapping agent is DNA. Table 3 shows that under conditions where DNA traps all the effluxed adriamycin, there is no progression to cell death, even with a very high intracellular concentration of drug (condition D). Thus, if drug is not available to interact with the cell surface, the cytotoxic response does not occur, no matter how much drug is present in the interior.

How can it be true both that intracellular DNA damage is required for cell death and that drug must interact with the surface membrane to cause cell death? This situation implies a communication between the membrane and the nucleus and suggests that adriamycin disruption of signal transduction systems may provide the mechanism by which cell death is initiated with this chemotherapeutic agent. It is notable that a similar idea is beginning to take hold in many of the laboratories studying cell death in the wide variety of other contexts discussed in this book.

Signaling Mechanisms in Adriamycin Action

All cells control their proliferation using a variety of growth factors and associated signal transduction mechanisms. These systems originate with signals at the plasma membrane and culminate in DNA replication and cell division. It would seem a reasonable idea that disruption of such events could be detrimental to a cell and could even lead to cessation of regulated metabolism and growth, i.e., to cell death. In fact, adriamycin and other drugs have been shown to disrupt a wide variety of signal transduction pathways, and we have reviewed this literature previously (Tritton and Hickman 1990). The best-understood example, or at least the one with the best-developed linkage to explaining the available facts, is the PKC pathway, and in the remainder of this paper, we summarize recent results from our laboratory.

It has been known for some time that adriamycin interacts with phospholipids, particularly acidic ones like cardiolipin (Duarte-Karim et al. 1976; Tritton et al. 1978; Murphree et al. 1982; Burke and Tritton 1985; Burke et al. 1988; Nicolay et al. 1988). Since binding to lipids perturbs membrane architecture, usually measured as fluidity changes, it seems reasonable that disruption of lipid structure could be an early initiating event in cytotoxicity. In fact, adriamycin does modulate membrane fluidity, and this has been suggested as an early event in cytotoxicity (Murphree et al. 1981; Sugiyama et al. 1986; Oth et al. 1987; Deliconstantinos et al. 1987; Lameh et al. 1989). Such a hypothesis is especially attractive in view of the previously discussed temperature data, since there is a solid↔fluid phase transition in the adriamycin-binding domain of the lipid bilayer at about the same temperature where cytotoxicity is switched on or off (~20°C). Membrane structure is not a function, however, only the milieu in which biological function takes place. Thus, the disruption of fluidity by adriamycin must be coupled to other functions, which in turn develop into cell death. One functional attribute of cell membranes that we have shown to be altered by adriamycin treatment is turnover of phosphoinositides (Posada et al. 1989a,b). In fact, the drug causes an increase in the turnover rate of this phospholipid (as well as phosphatidylcholine) and thus leads to an accumulation of the by-products of phospho-inositide turnover (diacylglycerol and inositol phosphates). These are not inert molecules but are intimately involved in growth regulation and other cellular responses: inositol triphosphate in the mobilization of intracellular Ca^{++}, and diacylglycerol by serving as an activator of PKC. The former subject is beyond the scope of this discussion, but the latter holds promise for explaining the cytotoxic action of adriamycin and for identifying the missing "factor" alluded to earlier, which may regulate the ability of drug to cause DNA damage.

If diacylglycerol increases in response to adriamycin, then this response should also lead to an increase in the activity of PKC in drug-exposed cells. This indeed occurs: The cytosolic activity of PKC is increased up to twofold by the drug without altering the activity of the membrane-bound form of the enzyme (Posada et al. 1989a). The next logical question then becomes, what is the consequence of PKC activation? Many proteins may serve as substrates for PKC, but the most important for this discussion is topoisomerase II. It was demonstrated several years ago (Sayhoun et al. 1986) that topoisomerase is phosphorylated by this Ser/Thr kinase and that phosphorylation controls the activity of DNA processing. We note that PKC also phosphorylates and controls the activity of topoisomerase I (Pommier et al. 1990), and other kinases may also act on each of the topoisomerases (e.g., casein kinase II; Ackerman et al. 1985), but we focus our discussion here on the topoisomerase II–PKC axis.

Since adriamycin treatment of cells affects the activity of PKC, it is possible that the reverse will be true as well, i.e., modulation of the activity of PKC may in turn modulate the activity of adriamycin. To explore this hypothesis, we turned to the phorbol ester TPA to alter PKC activity because this compound can both activate (by short treatment) and down-regulate (by extended treatment) the catalytic activity of PKC. Figure 2 shows that activation of PKC leads to an increase in topoisomerase-mediated DNA damage and an increase in cell death when cells are treated with adriamycin and, conversely, when PKC activity is reduced by down-regulation, there is less DNA damage and less cell death. Thus, there appears to be a regulated connection between the activity of PKC, the ability of

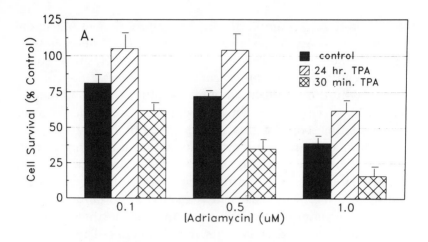

FIGURE 2 TPA effect on adriamycin cytotoxicity (A) and DNA damage (B). S180 cells were pretreated with 200 nM TPA for 30 min to activate PKC, or 24 hr to down-regulate PKC. The cells were then exposed to the indicated concentrations of adriamycin, and survival was determined by cloning in soft agar (A), or DNA damage was determined by alkaline elution (B). (Reprinted, with permission, from Posada et al. 1989a.)

topoisomerase II to damage DNA, and the response of the cell to die.

SUMMARY AND CONCLUSIONS

The results discussed here do not provide a definitive explanation of how adriamycin kills cancer (or any other) cells. They do, however, offer a framework for thinking about the process and for designing new experiments that may yield additional definition and clarity. It seems reasonably well established that the drug must interact with the plasma membrane to initiate cytotoxicity, but there is no concrete proof either way to establish the necessity for intracellular drug. One presumes

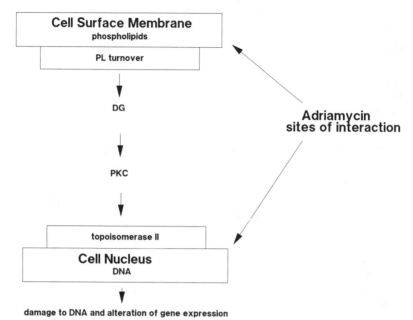

FIGURE 3 Proposed scheme by which adriamycin may induce cell death. The drug interacts with, and damages the function of, both the plasma membrane and nuclear DNA. Cytotoxicity does not ensue, however, in the absence of the membrane perturbation, and the activation of the PKC pathway is implicated in the signal transduction that needs to occur between the cell surface and the nucleus. (PL) Phospholipid, (DG) diacylglycerol, (PKC) protein kinase C. Details of the flow of events are described in the text.

nuclear drug is required as well, since there is evidence that the DNA helix must be distorted to cause topoisomerase II to malfunction (Liu 1989). The required membrane perturbation leads to the increased turnover of phospholipids that generates higher intracellular diglyceride; this is presumably catalyzed by phospholipase C, but whether the activity of this enzyme is directly controlled by interaction with adriamycin, or indirectly by alteration of membrane structure, has not been investigated. Diacyglycerol, in turn, activates PKC, and this protein may then increase the phosphorylation of many substrates, including topoisomerase II, which acts to cause lesions in DNA that ultimately lead to cell death (see Fig. 3). In this view, there is an ordered sequence of events, starting at the membrane and ending in the nucleus, which transduces the cytotoxic cascade. In effect then, there is a program of cell death, but we hesitate to use the term apoptosis because the generally accepted criteria for this process have not been met (or even systematically investigated) for adriamycin. Nonetheless, the existence of a particular ordered progression of events leading to the expiration of the cell suggests the possibility that there may be several pathways to cell death, and the one chosen depends on the type of insult. In addition to the events discussed here, adriamycin also perturbs other signaling pathways mentioned in various contexts in this volume, including cyclic nucleotides (Lehotay et al. 1982; Abraham et al. 1987), Ca^{++} regulation (Oakes et al. 1990), growth factor receptors (Zuckier and Tritton 1983), c-*fos* expression (Bhushan et al. 1991), and transglutaminase (Russell and Womble 1982). Because all of these signaling pathways act in the context of controlling specific gene expression, it is reasonable to suppose that adriamycin-induced cell death will share common elements with the gene-directed apoptotic pathways discussed by other contributors to this volume. Whatever the details, the final delineation of the pathway to cell death may involve complex interactions among the identified signaling and genetic players, or among new, as yet undiscovered, players.

ACKNOWLEDGMENTS

This work was supported in part by U.S. Public Health Service grant CA-44729 and American Cancer Society grant CH-392.

REFERENCES

Abraham, I., R.J. Hunter, K.E. Sampson, S. Smith, M.M. Gottesman, and J.K. Mayo. 1987. Cyclic AMP-dependent protein kinase regulates sensitivity of cells to multiple drugs. *Mol. Cell. Biol.* **7**: 3098.

Ackerman, P., C.V.C. Glover, and N. Osheroff. 1985. Phosphorylation of DNA topoisomerase II by casein kinase II: Modulation of eukaryotic topoisomerase II activity in vitro. *Proc. Natl. Acad. Sci.* **82**: 3164.

Arcamore, F. 1981. *Doxorubicin: Anticancer antibiotics.* Academic Press, New York.

Bachur, N.R., S.L. Gordon, M.V. Gee, and H. Kon. 1979. NADPH cytochrome P-450 reductase activation of quinone anticancer agents to free radicals. *Proc. Natl. Acad. Sci.* **76:** 954.

Burke, T.G. and T.R. Tritton. 1985. Structural basis of anthracycline sensitivity for unilamellar phosphatidylcholine vesicles: An equilibrium binding study. *Biochemistry* **24:** 1768.

Burke, T.G., A.C. Sartorelli, and T.R. Tritton. 1988. Selectivity of the anthracycline for negatively charged model membranes: Role of the amino group. *Cancer Chemother. Pharmacol.* **21:** 274.

Deliconstantinos, G., L. Kopeikina-Tsiboukidou, and V. Villiotou. 1987. Evaluation of membrane fluidity effects and enzyme activities alterations in adriamycin neurotoxicity. *Biochem. Pharmacol.* **36:** 1153.

Duarte-Karim, M., J.M. Ruysschaert, and J. Hildebrand. 1976. Affinity of adriamycin to phospholipids: A possible explanation for cardiac mitochondrial lesions. *Biochem. Biophys. Res. Commun.* **71:** 658.

Gianni, L., B.J. Corden, and C.E. Myers. 1983. The biochemical basis of anthracycline toxicity and antitumor action. *Rev. Biochem. Toxicol.* **5:** 1.

Kennedy, K.A., J.A. Siegfried, A.C. Sartorelli, and T.R. Tritton. 1983. The effects of anthracyclines on oxygenated and hypoxic tumor cells. *Cancer Res.* **43:** 54.

Lameh, J., R.Y. Chuang, M. Israel, and L.F. Chuang. 1989. Nucleoside uptake and membrane fluidity on N-trifluoroacetyladriamycin-14-O-hemiadipate-treated human leukemia and lymphoma cells. *Cancer Res.* **49:** 2905.

Lane, P., P. Vichi, D.L. Bain, and T.R. Tritton. 1987. Temperature dependence studies of Adriamycin uptake and cytotoxicity. *Cancer Res.* **47:** 4038.

Lehotay, D.C., B.A. Levey, B.J. Rogerson, and G.S. Levey. 1982. Inhibition of cardiac guanylate cyclase by doxorubicin and some of its analogs: Possible relationship to cardiotoxicity. *Cancer Treat. Rep.* **66:** 311.

Liu, L. 1989. DNA topoisomerase poisons as antitumor drugs. *Annu. Rev. Biochem.* **58:** 351.

Murphree, S.A., D. Murphy, A.C. Sartorelli, and T.R. Tritton. 1982. Adriamycin-liposome interactions: A magnetic resonance study of the differential effects of cardiolipin on drug-induced fusion and permeability. *Biochim. Biophys. Acta.* **691:** 97.

Murphree, S.A., T.R. Tritton, P.L. Smith, and A.C. Sartorelli. 1981. Adriamycin induced changes in the surface membrane of sarcoma 180 ascites cells. *Biochem. Biophys. Acta.* **649:** 317.

Nicolay, K., A.M. Sautereau, J.F. Tocanne, R. Brasseur, P. Huart, J.M. Ruysschaert, and B. deKruijff. 1988. A comparative model membrane study on structural effects of membrane-active positively charged anti-tumor drugs. *Biochem. Biophys. Acta.* **940:** 197.

Oakes, S.G., J.J. Schlager, K.S. Santone, R.T. Abraham, and G. Powis. 1990. Doxorubicin blocks the increase in intracellular Ca^{2+}, part of a second messenger system in WIE-115 murine neuroblastoma cells. *J. Pharmacol. Exp. Ther.* **252:** 979.

Oth, D., M. Begin, P. Bischoff, J.Y. Leroux, G. Mercier, and C. Bruneau. 1987. Induction, by adriamycin and mitomycin C, of modifications in lipid composition, size distribution, membrane fluidity and permeability of cultured RDM4 lymphoma cells. *Biochim. Biophys. Acta.* **900:** 198.

Pommier, Y., D. Kerrigan, K.D. Hartman, and R.I. Glazer. 1990. Phosphorylation of mammalian topoisomerase I and activation by protein kinase C. *J. Biol. Chem.* **265:** 9418.

Posada, J., P. Vichi, and T.R. Tritton. 1989a. Protein kinase C in Adriamycin action and resistance in mouse sarcoma 180 cells. *Cancer Res.* **49:** 6634.

Posada, J.A., E.M. McKeegan, K.F. Worthington, M.J. Morin, S. Jaken, and T.R. Tritton. 1989b. Human multidrug resistant KB cells overexpress protein kinase C: Involvement in drug resistance. *Cancer Comm.* **1:** 285.

Ross, W.E., D.L. Glaubiger, and K.W. Kohn. 1979. Qualitative and quantitative aspects of intercalator-induced DNA strand breaks. *Biochem. Biophys. Acta.* **5629:** 41.

Russell, D.H. and J.R. Womble. 1982. Transglutaminase may mediate certain physiological effects of endogenous amines and amine containing therapeutic agents. *Life Sci.* **30:** 1499.

Sayhoun, N., M. Wolf, J. Besterman, T. Hsieh, M. Sander, H. Levine, K. Chang, and P. Cuatrecasas. 1986. Protein kinase C phosphorylates topoisomerase II: Topoisomerase activation and its possible role in phorbol ester induced differentiation of HL-60 cells. *Proc. Natl. Acad. Sci.* **83:** 1603.

Sugiyama, M., T. Sakanashi, K. Okamoto, M. Chinami, T. Hidaka, and R. Ogura. 1986. Membrane fluidity in Ehrlich ascites tumor cells treated with adriamycin. *Biotechnol. Appl. Biochem.* **8:** 217.

Tewey, K.M., T.C. Rowe, L. Yand, B.D. Halligan, and L.F. Liu. 1984. Adriamycin-induced DNA damage mediated by mammalian DNA topoisomerase II. *Science* **226:** 466.

Tritton, T.R. 1991. Cell surface actions of adriamycin. *Pharmacol. Ther.* (in press).

Tritton, T.R. and J.A. Hickman. 1985. Cell surface membranes as a chemotherapeutic target. In *Experimental and clinical progress in cancer chemotherapy* (ed. F.M. Muggia), p. 81. Martinus Nijhoff, Boston.

——. 1990. How to kill cancer cells: Membranes and cell signalling as targets in cancer chemotherapy. *Cancer Cells* **2:** 95.

Tritton, T.R., S.A. Murphree, and A.C. Sartorelli. 1978. Adriamycin: A proposal on the specificity of drug action. *Biochem. Biophys. Res. Comm.* **84:** 802.

Vichi, P., S. Robison, and T.R. Tritton. 1989. Temperature dependence of adriamycin-induced DNA damage in L1210 cells. *Cancer Res.* **49:** 5575.

Zuckier, G. and T.R. Tritton. 1983. Adriamycin causes up regulation of epidermal growth factor receptors in actively growing cells. *Exp. Cell Res.* **148:** 155.

Zwelling, L.A., D. Kerrigan, and S. Michaels. 1982. Cytotoxicity and DNA strand breaks by 5-iminodaunorubicin in mouse leukemia L1210 cells: Comparison with Adriamycin and 4'-(9-acridinyl-amino)methanesulfon-m-aniside. *Cancer Res.* **42:** 2687.

Cellular and Molecular Aspects of Apoptosis in Experimental Tumors of Animals Treated with Analogs of LHRH and Somatostatin

B. Szende,[1,2] A.V. Schally,[1] A.M. Comaru-Schally,[1]
T.W. Redding,[1] G. Srkalovic,[1] K. Groot,[1] K. Lapis,[2] J. Timar,[2]
J. Neill,[3] and J. Mulchahey[3]

[1]Endocrine, Polypeptide and Cancer Institute, Veterans Administration Medical Center, New Orleans, Louisiana 70146 and Section of Experimental Medicine Department of Medicine, Tulane University Medical School, New Orleans Louisiana 70146
[2]First Institute of Pathology and Experimental Cancer Research, Semmelweis University Medical School, Budapest, Hungary
[3]Department of Physiology and Biophysics, University of Alabama at Birmingham, Alabama 35294

Since apoptosis (programmed cell death) was first described in 1972 (Kerr et al.), a series of studies have been carried out by endocrinologists, embryologists, and oncologists to elucidate the importance of this basic biological phenomenon in various fields of physiology and pathology (Kerr and Searle 1973; Wyllie et al. 1973a,b; Hopwood and Levison 1976; Ferguson and Anderson 1981; Wyllie and Morris 1982; Isaacs 1984; Morris et al. 1984; Wyllie 1985; Pierce et al. 1989; Kyprianou et al. 1990). In contrast to coagulation necrosis, apoptosis is considered to be an active cell function, namely suicide, when a programmed sequence of events leads to cell death (Haas 1989). According to Tomei et al. (1988), the activation of a so-called suicide gene is necessary to initiate apoptosis. After the activation of this gene, the synthesis of new RNA and protein species is induced (Wyllie et al. 1984). The activity of tissue transglutaminase (TG), an enzyme involved in the formulation of high-molecular-mass protein polymerase, becomes increased

(Fesus et al. 1987; Thomazy and Fesus 1989). This results in binding of intracytoplasmic membranous structures to the cell membrane, thus forming segregated parts of cytoplasm containing condensed cell organelles (Fesus et al. 1987). The endoplasmic reticulum becomes dilated, while the structures of mitochondria remain relatively preserved (Searle et al. 1975). A calcium-activated endogenous endonuclease is also presumed to be synthesized, which leads to characteristic DNA fragmentation (Wyllie 1980; Wyllie et al. 1981). Chromatin is first unevenly distributed within the nucleus and later condensed, leading to pycnosis or karyorrhexis (Searle et al. 1975; Wyllie et al. 1984). Fragmentation and lysis of the cytoplasm, together with the nuclear damage, are the morphological signs of programmed cell death (Kerr et al. 1972; Searle et al. 1975; Ferguson and Anderson 1981). Kerr and Harmon (this volume) describe in detail the morphological and biochemical features of apoptosis as well as the biological significance of this process.

The trigger of this process may be the deprivation of one or more of the growth factors or the down-regulation of growth factor receptors (Goustin et al. 1986; Haas 1989; Srkalovic et al. 1989). The possible causes and the morphological features of apoptosis in tumors were first described by Wyllie (1985, 1986) and Wyllie et al. (1980). Changes in hormonal environment and deprivation of growth factors were also mentioned by him (among others) as initiators of apoptosis. This type of programmed cell death occurs in various untreated tumors, especially in those that are hormone-dependent, such as some mammary carcinomas (Gullino 1980) and renal tumors (Bursch et al. 1988). Radiation-induced apoptosis was studied mostly in cell cultures (Tomei et al. 1988). Astonishingly little is known about cell death induced by hormonal therapy of tumors (Darbre et al. 1984; Taylor et al. 1984). Recently, Kyprianou et al. (1990) studied the mechanism of regression of prostatic cancer following castration using the androgen-responsive PC-82 human prostatic adenocarcinoma as a model system. They reported that the regression was due to biochemical and morphological events that resulted in the cessation of cell proliferation and activation of apoptosis. To the best of our knowledge, no histological studies were carried out on the effect on hormones, especially peptide hormones, on the apoptotic activity of tumor tissue.

The antitumor effect of luteinizing hormone-releasing hormone (LHRH) and somatostatin analogs has been widely studied by our group (see, e.g., Schally and Comaru-Schally 1987; Schally 1988). The regressive changes in the treated tumors (Paz-Bouza et al. 1987; Zalatnai and Schally 1989) were evaluated by us in detail and classified as apoptosis or as the results of apoptosis (Szende et al. 1989a,b; 1990a,b,c).

In this paper, our observations on the induction of apoptosis by LHRH and somatostatin analogs on various experimental animal tumors are summarized, and the possible mode of action of these peptides is discussed. This study is based on our histological, ultrastructural, and immunohistochemical observations, obtained by the reevaluation of our previously published experimental material (Redding and Schally 1983a,b; Zalatnai et al. 1988; Szende et al. 1989a,b) or acquired in recently performed experiments (Szende et al. 1990a,b,c).

EXPERIMENTAL DATA

The peptides and sustained release preparations used in these studies were as follows:

Microcapsule formulation of the LHRH agonist D-Trp-6-LHRH (pyro-Glu-His-Trp-Ser-Tyr-D-Trp-Leu-Arg-Pro-Gly-NH$_2$) in biodegradable poly-(DL-lactide-coglycolide) was prepared by Dr. P. Orsolini (Cytotech, Martigny, Switzerland) and provided by Debiopharm SA (Lausanne, Switzerland). Various batches of microcapsules of the somatostatin analog D-Phe-Cys-Tyr-D-Trp-Lys-Val-Cys-Trp-NH$_2$ (RC-160) (Novabiochem, Laufelfingen, Switzerland) in poly-(DL-lactide-coglycolide) (Cytotech, Martigny, Switzerland) were also designed to release 15 µg per day and 48 µg per day, respectively.

The LHRH antagonist [Ac-D-Nal(2)', D-Phe(4Cl)^2, D-Pal(3)^3, D-Cit^6, D-Ala^{10}]LHRH (SB-75) (ASTA Pharma Co., Germany) was administered either in osmotic minipumps releasing 50 µg per day (to mice) or in the form of microcapsules in poly-(DL-lactide-coglycolide) liberating 8–25 µg per day for 8 or 21 days (to mice and hamsters) or 71 µg per day to rats. Pancreatic carcinomas were induced by the administration of N-(nitro-sobis-2-oxopropyl)amine (BOP) (Szende et al. 1989b, 1990b,c; Zalatnai and Schally 1989). Estrogen-receptor-positive MXT mouse mammary carcinoma (Redding and Schally 1983a; Szende et al. 1989a, 1990a), androgen-dependent R3327H Dunning rat prostate adenocarcinoma (Zalatnai et al. 1988), and Swarm rat chondrosarcoma (Redding and Schally 1983b) were transplanted subcutaneously.

The treatment of hamsters with the analogs was initiated 24 weeks after the first administration of the carcinogen and lasted for 8 weeks (Szende et al. 1989b, 1990b,c; Zalatnai and Schally 1989). The transplantable mouse and rat tumors were first treated when the tumors reached a measurable volume. The treatment lasted for 21 or 30 days (Redding and Schally 1983a,b; Zalatnai et al. 1988; Szende et al. 1989a, 1990a).

Histological, ultrastructural, and immunohistochemical studies were carried out on the tumors. Biochemical radiore-ceptor assays for D-Trp-6-LHRH, somatostatin, epidermal growth factor (EGF), and insulin-like growth factor I (IGF-I) and the determination of serum luteinizing hormone (LH), growth hormone (GH), IGF-I, and estrogen levels were also performed.

The anti-LHRH receptor antibody was polyclonal, prepared by immunization of a single rabbit with high-performance liquid chromatography, PAGE-purified rat LHRH receptor (Mulchahey and Neill 1989). Purified immunoglobulins were prepared from three pooled bleedings by chromatography on Protein A–Sepharose. Working dilutions of immunoglobulins were prepared from the resulting stock which was 4.3 mg protein/ml. These immunoglobulins were determined to recognize the LHRH receptor by criteria identical to those used by Mulchahey et al. (1986) in evaluating other anti-LHRH receptor immunoglobulins.

For electronmicroscopic immunohistochemistry directed to LHRH receptors, 50-mm sections were cut from the blocks embedded into Spurr's low viscosity medium and mounted on gold grids. The immunogold method of Roth (1983) was used. Deletion of the primary antibody was used as a negative control.

Response of Hormone-dependent Tumors to LHRH and Somatostatin Analogs

Tumor growth. The treatment with LHRH and somatostatin analogs resulted in significant inhibition of tumor growth in the case of the MXT mouse mammary carcinoma (D-Trp-6-LHRH, RC-160, SB-75) (Redding and Schally 1983a; Szende et al. 1989a, 1990a), Dunning prostate carcinoma of rats (D-Trp-6-LHRH, RC-160, SB-75) (Schally and Redding 1987; Zalatnai et al. 1988), and Swarm rat chondrosarcoma (D-Trp-6-LHRH) (Redding and Schally 1983b). Table 1 summarizes some of these effects, indicating the growth inhibition in percentage of the controls. In some of the experiments, the combination of RC-160 and D-Trp-6-LHRH was used, and this treatment appeared to be more effective than either D-Trp-6-LHRH or RC-160 alone.

Serum LH, GH, IGF-I, and estradiol levels. Chronic treatment with D-Trp-6-LHRH or with SB-75, lasting 1–2 months, resulted in a decrease in serum LH and estradiol levels. Serum GH was lowered by somatostatin analog RC-160. IGF-I levels were moderately decreased after the treatment with the LHRH agonists and antagonists and somatostatin analogs (Schally and Redding 1987; Szende et al. 1990a,c).

Receptor responses. Membrane binding sites for LHRH and somatostatin were found in different experimental tumors, including MXT mouse mammary cancer and BOP-induced pancreatic tumor of hamsters. The affinity constant and the binding capacity of these receptors decreased significantly after treatment with D-Trp-6-LHRH, SB-75, or RC-160 (Fekete et al. 1989; Srkalovic et al. 1989; Szende et al. 1990a,b,c). Receptors for EGF and IGF-I were also detected in these tumors. Binding capacity of IGF-I receptors in MXT mammary tumors was reduced after prolonged treatment with D-Trp-6-LHRH and RC-160 (Srkalovic et al. 1989). Microcapsules of SB-75 and D-Trp-6-LHRH also produced down-regulation of EGF receptors. LHRH receptors were also identified by immuno-electron microscopy using the immunogold technique. Cells of untreated mouse MXT mammary carcinoma and untreated BOP-induced hamster pancreatic carcinoma (HPC) showed

TABLE 1 INHIBITORY EFFECT OF D-TRP-6-LHRH, RC-160, AND SB-75 ON MXT MOUSE MAMMARY TUMOR, DUNNING RAT PROSTATE TUMOR, SWARM RAT CHONDROSARCOMA, AND BOP-INDUCED HAMSTER PANCREATIC CARCINOMA

Treatment	Tumor	Daily dose (μg)	Inhibition of tumor weight (%)	Reference
D-Trp-6-LHRH	MXT[a]	25	63	Szende et al. (1989a)
D-Trp-6-LHRH	HPC[b]	25	55	Zalatnai and Schally (1989)
D-Trp-6-LHRH	Swarm[c]	25	40	Redding and Shally (1983b)
D-Trp-6-LHRH	Dunning[c]	25	73	Zalatnai et al. (1988)
RC-160	MXT[a]	25	44	Szende et al. (1989a)
RC-160	HPC[b]	15	35	Szende et al. (1989b)
RC-160	HPC[b]	25	57	Zalatnai and Schally (1989)
RC-160	HPC[b]	48	72	Szende et al. (1990c)
RC-160	Dunning[c]	5	58	Zalatnai et al. (1988)
D-Trp-6-LHRH plus RC-160	MXT[a]	25 + 25	62	Szende et al. (1989a)
D-Trp-6-LHRH plus RC-160	HPC[b]	25 + 48	85	Szende et al. (1990c)
D-Trp-6-LHRH plus RC-160	Dunning[c]	25 + 5	84	Zalatnai et al. (1988)
SB-75	MXT[a]	25	84	Szende et al. (1990a)
SB-75	HPC[b]	8	70	Szende et al. (1990b)

Duration of treatment: [a] Three weeks; [b] two months; [c] one month. Treatment was performed by the injection of microcapsules that released the daily amount of peptides, indicated in the table. In the case of Swarm tumor, daily injections were given. The significance of published data is reported in the references listed.

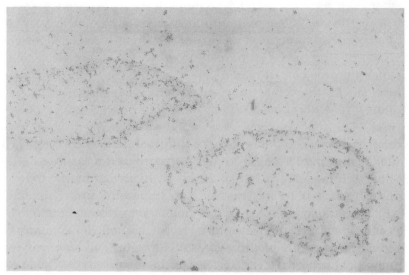

FIGURE 1 LHRH receptor visualization in BOP-induced HPC cells. The nuclei are heavily labeled by G 10 colloidal gold particles (anti LHRH/ 1:1,000/-GAR G 10/1:20 immunogold method). Magnification, 5200x.

positivity for LHRH receptors. The ratio of positive cells was ~25% in the MXT tumor and ~80% in the pancreatic cancer. The labeling was found in both kinds of tumor cells, predominantly in the nuclei (Fig. 1).

Histology. The regressive changes in the tumors of animals treated with the analogs have been described previously (Paz-Bouza et al. 1987; Zalatnai et al. 1988; Zalatnai and Schally 1989). After we discovered that the basic change in tumor tissue following hormonal treatment is, in fact, the enhancement of apoptosis (programmed cell death), we carried out a systematic histological work to reevaluate the material of our previous experiments and to add new histological samples by additional treatment of tumor-bearing animals as described above. The results of our histological studies are summarized below.

The early signs of apoptosis are characterized by the marginization or uneven distribution of the chromatin in the nucleus and the swelling of the cytoplasm. This is followed by either pycnosis or fragmentation of the nucleus (Fig. 2). The cytoplasm becomes strongly acidophilic, and, later on, lysis or fragmentation takes place. In the case of adenocarcinomas (pancreatic, mammary, and prostatic), this process can be observed in single cells or more often in several neighboring cells, which become desquamated and can be detected in the acinar lumina. At the same time, mesenchymal cells, like macrophages and fibroblasts, enter the tumorous glandular structures. The macrophages clear up the fragments of apoptotic cells. The fibroblasts become the lining of the tumorous glandular structures. The end stage of this process is the persistence of lumina lined by flat fibrocytes, surrounded by fibrous or scar tissue (Fig. 3). Thus, the skeleton of the tumor remains without tumor cells. In some cases, the destruction of the tumorous glandular structures is accompanied by a strong mononuclear inflammatory infiltrate. Again, in some of the glandular structures that undergo apoptosis, the calcification of apoptotic cells, which can become extensive, can be ob-

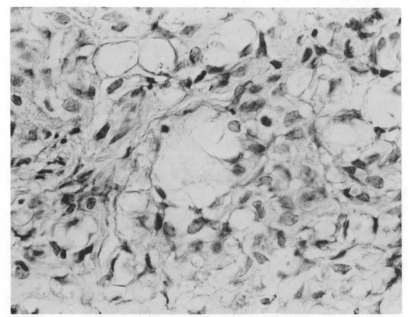

FIGURE 2 Swelling of the cytoplasm and nuclear pycnosis in several tumor cells of HPC after SB-75 treatment (H and E). Magnification, 450x.

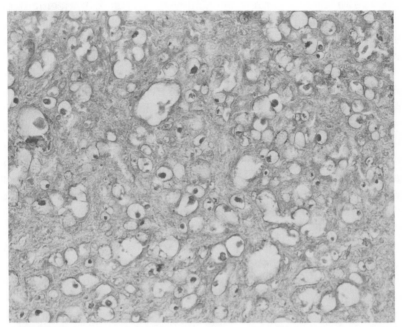

FIGURE 3 Total replacement of tumor cells in glandular spaces by mesenchymal cells. Note scar formation around the spaces. HPC, treated with SB-75 (H and E). Magnification, 288x.

served. In some cases, metaplastic ossification occurs in the newly formed connective tissue among the apoptotic glandular structures. If neighboring glandular spaces become destroyed by apoptosis, these structures may melt into small foci, which cannot be distinguished from coagulation necrosis. Occasionally, coagulative necrotic areas can be found in the center of the tumors, presumably due to hypoxia. The process of apoptosis can be detected in both untreated and treated tumors. To quantitate this phenomenon, we introduced the term "apoptotic index," which indicates the percentage of tumorous glandular structures showing any sign of apoptosis, compared to the total number of the glandular structures in the tumor tissue. In control, untreated animals, this ratio is usually under 1–2%. After treatment with the analogs, this ratio becomes much higher: as high as 40%, and after a very high dose, even 86–89%.

This degree of apoptosis signifies that total destruction of the tumor tissue cannot be achieved by the enhancement of the programmed cell death. Some parts of the tumor tissue seem to be sensitive to the treatment, and some seem to be resistant. Mitotic activity is present in the nonaffected parts of the tumors, the ratio of which is between 1% and 3%, according to the type and age of the tumor. The growth of the tumor as a whole depends on the ratio of cell loss produced by apoptosis as well as coagulation necrosis, and on the cell gain resulting from mitosis. Combination treatment with both D-Trp-6-LHRH and RC-160, as well as therapy with SB-75, causes an adequate degree of apoptosis, which, together with other processes responsible for inhibition of proliferation, leads to the cessation of tumor growth or even regression of tumors.

The process of apoptosis can also be observed in the chondrosarcoma cells. Because these cells are surrounded by the cartilage matrix, the elimination of the dead tumor cells by macrophages is virtually impossible. Such dead tumor cells with pycnotic nuclei and swollen or lytic cytoplasm can be seen scattered in the periphery of the tumor tissue among viable and even dividing tumor cells. The frequency of apoptotic cells after D-Trp-6-LHRH treatment was 15–20% of the total number of tumor cells, in contrast to the 1–2% control value. Large central areas of coagulation necrosis were seen in the control and treated tumors.

Apoptotic cells could also be studied under the electron microscope in the case of the MXT mouse mammary tumor and BOP-induced HPC. The marginization of chromatin and pycnosis was the most common finding in the nuclei (Fig. 4). In the cytoplasm, the endoplasmic reticulum showed cistern-like dilatation in some of the apoptotic cells. Segregation of parts of the cytoplasm containing relatively intact organelles was another important finding. Myelin figures and phagosomes occurred in the cytoplasm of tumor cells undergoing apoptosis. In contrast to coagulation necrosis, apoptosis was confined to individual tumor cells or small groups of neighboring tumor cells. Apoptosis never resulted in a mass of disorganized cell debris, as seen in coagulation necrosis. The phagocytic activity of macrophages ensured the rapid clearing of the remnants of apoptotic cells.

TG activity was seen in few cells of control tumors and in

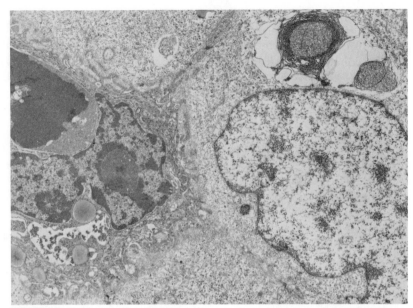

FIGURE 4 Tumor cell with pycnotic nucleus (*right*) and a cell with an apoptotic body (*left*) in an MXT tumor treated with SB-75. Magnification, 4940x.

an increased number of cells in MXT mammary tumor tissue of the animals treated with the analogs (Fig. 5). In some of these cells, the TG activity was confined to apoptotic bodies, but in others, no such correlation could be observed.

IGF-I receptors were detected by immunohistological methods both in the MXT mammary adenocarcinoma and in the BOP-induced hamster pancreatic adenocarcinoma cells. The estimated number of these receptors decreased after the treatment with D-Trp-6-LHRH or RC-160. Therapy with SB-75 or D-Trp-6-LHRH decreased the estimated number of EGF receptors (Szende et al. 1990a,b,c). Binding of D-Trp-6-LHRH and RC-160 to several pancreatic adenocarcinoma cells could be demonstrated immunohistochemically after the treatment with these peptides (Szende et al. 1990c).

DISCUSSION

Apoptosis is a basic biological phenomenon that occurs in normal tissues and in tumors as well (Kerr et al. 1972; Wyllie et

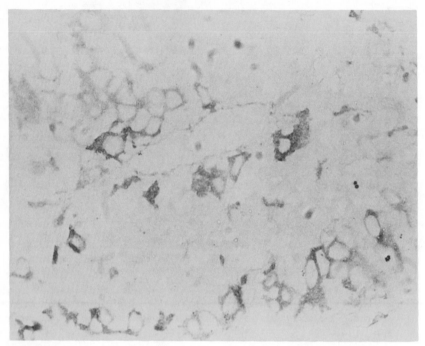

FIGURE 5 Tissue TG positivity in the cytoplasm of several MXT tumor cells after RC-160 treatment (immunoperoxidase). Magnification, 475x.

al. 1973a,b). Apoptosis can be enhanced by various manipulations in tumors and in preblastomatous lesions (Bursch et al. 1984; Tomei et al. 1988). The results of our studies show that in vivo treatment with agonistic and antagonistic analogs of LHRH and somatostatin analog RC-160 causes a significant enhancement of apoptosis in hormone-sensitive tumors. In present oncological terminology, the hormone sensitivity usually means only the sensitivity to sex steroids, but in the future, this definition should extend also to analogs of hypothalamic peptides and other hormones. Some of the inhibitory effects of LHRH analogs and somatostatin analogs on tumors may not be mediated by a fall in sex steroid levels or GH-IGF-I levels, respectively. Thus, tumors with special binding sites for LHRH and its analogs and for somatostatin, such as BOP-induced pancreatic carcinoma of hamsters and MXT mouse mammary tumor, may respond directly to peptide treatment with inhibition of growth. The fact that LHRH receptors were found by us also morphologically in experimental mammary and pancreatic cancer cells emphasizes the important role of these receptors in the therapeutic responses of these tumors to LHRH analogs. The dephosphorylation of EGF receptors by hormonal treatment can also be responsible for tumor growth inhibition (Liebow et al. 1989). The mechanism by which the action of peptides on various receptors is transferred to the nucleus is still under investigation (Wyllie et al. 1984; Wyllie 1985, 1986; Tomei et al. 1988; Haas 1989; Trauth et al. 1989). According to our findings, the action of the hypothalamic hormone analogs is not followed by a blockade or reduction in number of mitotic cells. It may be assumed that the expression of a specific gene (Tomei et al. 1988) is triggered. This "suicide gene" is assumed to produce new RNA and protein species, the effect of which results in a programmed cell death. The increased number of apoptotic cells after treatment with the LHRH and somatostatin analogs strongly supports this hypothesis.

Increased TG activity after the administration of the analogs points to a close relationship between apoptosis and the activation of TG (Fesus et al. 1987; Thomazy and Fesus 1989). The function of TG is directed toward the repair of cytoplasmic damage during the course of apoptosis by formation of apoptotic bodies, i.e., segregates of damaged cell organelles. We assume that TG expression is part of the apoptotic process. Together with the apoptotic index, the percentage of TG-positive tumor cells might be used as an indicator of responsiveness of tumors to hormonal therapy.

In conclusion, according to our findings, LHRH and somatostatin analogs act on tumor growth by enhancing apoptosis. Enhanced apoptosis results in an increased cell loss, which competes with the cell gain produced by mitotic activity. As a result of this competition, tumor growth is slowed down. This mechanism of action of hypothalamic analogs differs completely from that of cytostatic agents and is directed against nondividing cells. LHRH and somatostatin analogs should be considered to be useful agents in combined modality treatment of hormone-sensitive tumors.

ACKNOWLEDGMENTS

Some experimental work described in this paper was supported by U.S. Public Health Service grants CA-40003, CA-40077, and CA-40004 and by the Medical Research Service of the Department of Veterans Affairs.

REFERENCES

Bursch, W., J. Leihr, D. Sirbasku, and R. Schulte-Hermann. 1988. The role of cell death for growth and regression of hormone-dependent H-301 hamster kidney tumors. In *Chemical carcinogenesis. Models and mechanisms* (ed. F. Feo et al.), p. 275. Plenum Press, New York.

Bursch, W., B. Lauer, I. Timmerman-Trosiener, G. Barthel, J. Schuppler, and R. Schulte-Hermann. 1984. Controlled death (apoptosis) of normal and putative preneoplastic cells in rat liver following withdrawal of tumor promoters. *Carcinogenesis* **5:** 543.

Darbre, P.D., S. Curtis, and R.J. King. 1984. Effects of estradiol and tamoxifen on human breast cancer cells in serum free culture. *Cancer Res.* **44:** 2790.

Fekete, M., A. Zalatnai, and A.V. Schally. 1989. Presence of membrane binding sites for D-Trp-6-luteinizing hormone releasing hormone in experimental pancreatic cancer. *Cancer Lett.* **45:** 87.

Ferguson, D.J.P. and T.J. Anderson. 1981. Ultrastructural observations on cell death by apoptosis in the "resting" human breast. *Virchows Arch. Pathol. Anat.* **393:** 193.

Fesus, L., V. Thomazy, and A. Falus. 1987. Induction and activation of tissue transglutaminase during programmed cell death. *FEBS Lett.* **224:** 104.

Goustin, A.S., E.B. Leof, G.D. Shipley, and H.L. Moses. 1986. Growth factors and cancer. *Cancer Res.* **46:** 1015.

Gullino, P.M. 1980. The regression process in hormone-dependent mammary-carcinomas. In *Hormones and cancer* (ed. S. Jacobelli et al.), p. 271. Raven Press, New York.

Haas, R. 1989. For whom the cell tolls: A portrait of physiological cell death. *J. Natl. Inst. Health* **1:** 91.

Hopwood, D. and D.A. Levison. 1976. Atrophy and apoptosis in the cyclical human endometrium. *J. Pathol.* **11:** 159.

Isaacs, J.T. 1984. Antagonistic effect of androgen on prostatic cell death. *Prostate* **5:** 545.

Kerr, J.F.R. and J. Searle. 1973. Deletion of cells by apoptosis during castration-induced involution of rat prostate. *Virchows Arch. Abt. B.* **13:** 87.

Kerr, J.F.R., A.H. Wyllie, and A.R. Currie. 1972. Apoptosis: A basic biological phenomenon with wide-ranging implications in tissue kinetics. *Br. J. Cancer* **26:** 239.

Kyprianou, N., H.F. English, and J.T. Isaacs. 1990. Programmed cell death during regression of PC-82 human prostate cancer following androgen ablation. *Cancer Res.* **50:** 3748.

Liebow, C., C. Reilly, M. Serrano, and A.V. Schally. 1989. Somatostatin analogues inhibit growth of pancreatic cancer by stimulating tyrosine phosphatase. *Proc. Natl. Acad. Sci.* **86:** 2003.

Morris, R.G., E. Duvall, A.D. Hargreaves, and A.H. Wyllie. 1984. Hormone-induced cell death 2. Changes in the cell surfaces of apoptotic thymocytes. *Am. J. Pathol.* **115:** 426.

Mulchahey, J.J. and J.D. Neill. 1989. Production and characterization of polyclonal antibodies to the GnRH receptor. In *Abstract 941 from the Endocrine Society Meeting*, Seattle, Washington, p. 258.

Mulchahey, J.J., J.D. Neill, L.D. Dion, K.L. Bost, and J.E. Blalock. 1986. Antibodies to the binding site of the receptor for LHRH: Generation with a synthetic decapeptide encoded by an RNA complementary to LHRH mRNA. *Proc. Natl. Acad. Sci.* **83:** 9714.

Paz-Bouza, J.I., T.W. Redding, and A.V. Schally. 1987. Treatment of nitrosamine-induced pancreatic tumors in hamsters with analogs of somatostatin and luteinizing hormone-releasing hormone. *Proc. Natl. Acad. Sci.* **84:** 1112.

Pierce, G.B., A.L. Lewellyn, and R.E. Parchment. 1989. Mechanism of programmed cell death in the blastocyst. *Proc. Natl. Acad. Sci.* **86:** 3654.

Redding, T.W. and A.V. Schally. 1983a. Inhibition of mammary tumor growth in rats and mice by administration of agonistic analogs of luteinizing hormone-releasing hormone. *Proc. Natl. Acad. Sci.* **80:** 1459.

———. 1983b. Inhibition of growth of the transplantable rat chondrosarcoma by analog of hypothalamic hormones. *Proc. Natl. Acad. Sci.* **80:** 1078.

Roth, J. 1983. The colloidal gold marker system for light and electron microscopic cytochemistry. In *Techniques in immunocytochemistry* (ed. G.R. Bullock and P. Petrusz), p. 217. Academic Press, New York.

Schally, A.V. 1988. Oncological applications of somatostatin analogs. *Cancer Res.* **48:** 6977.

Schally, A.V. and A.M. Comaru-Schally. 1987. Use of luteinizing hormone-releasing hormone analogs in the treatment of hormone-dependent tumors. *Semin. Reprod. Endocrinol.* **5:** 386.

Schally, A.V. and T.W. Redding. 1987. Somatostatin analogs as adjuncts to agonists of luteinizing hormone-releasing hormone in the treatment of experimental prostatic cancer. *Proc. Natl. Acad. Sci.* **84:** 7275.

Searle, J., T.A. Lawson, P.J. Abbott, B. Harmon, and J.F.R. Kerr. 1975. An electron-microscopic study of the mode of cell death induced by cancer-chemotherapeutic agents in populations of proliferating normal and neoplastic cells. *J. Pathol.* **116:** 129.

Srkalovic, G., B. Szende, T.W. Redding, K. Groot, and A.V. Schally. 1989. Receptors for D-Trp-6-luteinizing hormone-releasing hormone, somatostatin and insulin-like growth factor I in MXT mouse mammary carcinoma. *Proc. Soc. Exp. Biol. Med.* **192:** 209

Szende, B., A. Zalatnai, and A.V. Schally. 1989a. Programmed cell death (apoptosis) in pancreatic cancers of hamsters after treatment with analogs of both luteinizing hormone-releasing hormone and somatostatin. *Proc. Natl. Acad. Sci.* **86:** 1643.

Szende, B., K. Lapis, T.W. Redding, G. Srkalovic, and A.V. Schally. 1989b. Growth inhibition of MXT mammary carcinoma by enhancing programmed cell death (apoptosis) with analogs of both luteinizing-releasing hormone and somatostatin. *Breast. Cancer Res. Treat.* **14:** 307.

Szende, B., G. Srkalovic, K. Groot, K. Lapis, and A.V. Schally. 1990a. Growth inhibition of mouse MXT mammary tumor by the luteinizing hormone-releasing hormone antagonist SB-75. *J. Natl. Cancer Inst.* **82:** 513.

———. 1990b. Regression of nitrosamine-induced pancreatic cancers in hamsters treated with luteinizing hormone-releasing hormone antagonists or agonists. *Cancer Res.* **50:** 3716.

Szende, B., G. Srkalovic, A.V. Schally, K. Lapis, and K. Groot. 1990c. Inhibitory effects of analogs of luteinizing hormone-releasing hormone and somatostatin on pancreatic cancers in hamsters. *Cancer* **65:** 2279.

Taylor, C.M., B. Blanchard, and D.T. Zava. 1984. Estrogen receptor-mediated and cytotoxic effects of antiestrogens tamoxifen and 4-hydroxy-tamoxifen. *Cancer Res.* **44:** 1409.

Thomazy, V. and L. Fesus. 1989. Differential expression of tissue transglutaminase in human cells. An immunohistochemical study. *Cell Tissue Res.* **255:** 215.

Tomei, L.D., P. Kanter, and C.E. Wenner. 1988. Inhibition of radiation-induced apoptosis in vitro by tumor promoters. *Biochem. Biophys. Res. Commun.* **155:** 324.

Trauth, B.C., C. Klas, A.M.J. Peters, S. Matzku, P. Moller, W. Falk, K.M. Debatin, and P.H. Krammer. 1989. Monoclonal antibody-mediated tumor regression by induction of apoptosis. *Science* **245:** 301.

Wyllie, A.H. 1980. Glucocorticoid-induced thymocyte apoptosis is associated with endogenous endonuclease activation. *Nature* **284:** 555.

———. 1985. The biology of cell death in tumours. *Anticancer Res.* **5:** 131.

———. 1986. What is apoptosis? *Histopathology* **10:** 995.

Wyllie, A.H. and R.G. Morris. 1982. Hormone-induced cell death. Purification and properties of thymocytes undergoing apoptosis after glucocorticoid treatment. *Am. J. Pathol.* **109:** 78.

Wyllie, A.H., G.J. Beattie, and A.D. Hargreaves. 1981. Chromatin changes in apoptosis. *Histochem. J.* **13:** 681.

Wyllie, A.H., J.F.R. Kerr, and A.R. Currie. 1973a. Cell death in the normal neonatal rat adrenal cortex. *J. Pathol.* **111:** 225.

————. 1980. Cell death: The significance of apoptosis. *Int. Rev. Cytol.* **68:** 251.

Wyllie, A.H., J.F.R. Kerr, I.A.M. Macaskill, and A.R. Currie. 1973b. Adrenocortical cell deletion: The role of ACTH. *J. Pathol.* **111:** 85.

Wyllie, A.H., R.G. Morris, A.L. Smith, and D. Dunlop. 1984. Chromatin cleavage in apoptosis: Association with condensed chromatin morphology and dependence on macromolecular synthesis. *J. Pathol.* **142:** 66.

Zalatnai, A. and A.V. Schally. 1989. Treatment of the *N*-nitrosobis (2 oxopropyl)amine-induced pancreatic cancer in Syrian golden hamsters with D-Trp-6-LH-RH and a somatostatin analog RC-160. *Cancer Res.* **49:** 1810.

Zalatnai, A., J.I. Paz-Bouza, T.W. Redding, and A.V. Schally. 1988. Histologic changes in the rat prostate cancer model after treatment with somatostatin analogs and D-Trp-6-LH-RH. *Prostate* **12:** 85.

Genetic Response of Prostate Cells to Androgen Deprivation: Insights into the Cellular Mechanism of Apoptosis

R. Buttyan
Department of Urology, Columbia University
New York, New York 10032

The prostate gland is a male sexual accessory organ positioned at the base of the mammalian bladder. During ejaculation, this gland secretes large amounts of proteinaceous substance that is added to the seminal fluid. The value of this contribution to male reproduction is far from established. However, because of the disproportionate level to which growth aberrations of the prostate cause health problems during human aging, biomedical science has focused some degree of attention on defining the agents that support the development and growth of this organ. At the top of this list are the male sex steroids (androgens), represented principally by testosterone.

Since testosterone is produced by the testis, it is easy to understand the rationale of attempts to treat both benign and malignant prostate conditions by castrations. This approach was in use as early as the late 19th century and, by 1941, Huggins and his colleagues had firmly established the usefulness of castration as a therapy for human metastatic prostate cancer (Huggins and Hodges 1941). This treatment is still common today and provides symptomatic relief for most prostate cancer patients because the prostate gland, the primary tumors, and their metastasis shrink (i.e., involute). It was apparent early on that the survival of both normal and malignant prostate cells was dependent on androgenic steroids; however, the basis for this dependence remained un-

Apoptosis: The Molecular Basis of Cell Death
Copyright 1991 Cold Spring Harbor Laboratory Press 0-87969-366-5/91 $3.00 + 00

clear. Although it was known that androgen deprivation could decrease the rate of DNA synthesis in the gland (Sufrin and Coffey 1973; Coffey et al. 1976), this alone could not explain the rapid structural and cytological changes that accompanied the involution of the prostate following castration.

In 1975, Bruchovsky and his colleagues made a unique and important observation concerning the potential nature of the relationship between androgen withdrawal and prostatic involution. They reported that castration was followed by the rapid onset of cell death in the normal prostate gland (Bruchovsky et al. 1975). Cell death was so massive that in the model they studied, the rat ventral prostate gland, up to 85% of the cells were eliminated from the gland within a week of castration (Lesser and Bruchovsky 1973). Moreover, the physical characteristics of the dying prostate cells led them to suggest that the cells were actively killing themselves. These investigators called this active process of prostate cell death an "autophagia." Since then, many other investigators have confirmed their observations (Lee 1981). Mathematical modeling studies have substantiated drastic elevations in the rate of cell death in the rat ventral prostate gland (Isaacs 1984). Histological studies have further demonstrated that the secretory epithelial cells, the overwhelming majority of cells in the ventral prostate, are virtually eliminated after castration through the process of apoptosis (Kerr and Searle 1973; Sanford et al. 1984; English et al. 1985, 1989). In view of the medical importance of these observations, the ability to induce extensive cellular apoptosis in the normal prostate gland has resulted in extensive development of this system as a general model for apoptosis. Recent application of molecular analysis to this model system has provided important new insights into the cellular basis of apoptosis and gives promise to our ultimate ability to decipher the molecular mechanism of this physiological process.

The Dying Prostate Cell

Castration initiates a massive glandular remodeling process in the rat ventral prostate. This process begins by 12 hours, when the nuclear androgen receptor content of the prostatic cells declines to undetectable levels (Kyprianou and Isaacs 1988). The activity of a Ca^{++}, Mg^{++}-dependent endonuclease is detectably elevated in extracts of ventral prostate nuclei within 24 hours of castration. This appears to be coordinated with an increase in the fragmentation of prostatic nuclear DNA (English et al. 1989). By 24 hours after castration, epithelial cells, especially those of the prostatic acini, show a decline in columnar height (Kiplesund et al. 1988) and a slight increase in the presence of small, vacuolated, dark-staining cells throughout the acini that can be characterized as apoptotic bodies. The number of apoptotic bodies increases rapidly thereafter and reaches a maximum at 2–3 days following castration (Sanford et al. 1984; English et al. 1989). These apoptotic bodies are ultimately removed from the tissue by the phagocytic action of neighboring cells in addition to local macrophages (Helminen and Ericsson 1972). Approximately 3 days after castration, macrophages from outside the prostate are transiently recruited to the area in response to the presence of large numbers of apoptotic cells (Helminen and Ericsson 1972; English et al. 1989), which accounts for the swift decline in the number of apoptotic bodies seen after this time. The synthesis of prostatic secretory proteins decreases drastically during the involution period, coincident with the loss of the secretory epithelial cells (Parker and Scrace 1979).

The vestigial ventral prostate tissue that remains 1 week after castration consists mainly of stromal fibroblasts; however, the lumens of the glandular areas that remain are populated by small cuboidal epithelial cells which are apparently resistant to the androgen-deprived environment. Of further interest, the normal prostatic glandular structure in the castrate rat can be rapidly regenerated by returning physiological levels of testosterone. The gland is repopulated by epithelial cells that are as dependent on testosterone as those of the progenitor adult organ; removal of the testosterone source will just as rapidly reinitiate the involution response as did the original castration event (Sanford et al. 1984).

Androgen-deprived rat ventral prostate epithelial cells exhibit numerous attributes that are now known to be characteristic of cells undergoing apoptosis (Kerr and Searle 1973; Sanford et al. 1984; English et al. 1989). Microscopic analysis of thin sections through involuting rat ventral prostate glands obtained at 3 days after castration reveals a large number of shrunken epithelial cells having condensed, irregular, and fragmented nuclei typical for apoptotic bodies (Fig. 1). High-molecular-weight DNA extracted from this tissue demonstrates the apoptosis-associated "ladder" pattern following agarose gel electrophoresis (Kyprianou and Isaacs 1988), in contrast to the distinct high-molecular-weight band of DNA extractable from the normal mature prostate gland (Fig. 2). This ladder results from the presence of DNA fragments having sizes in multiples of 180-bp units and is thought to occur because of endonuclease action specific to the internucleosomal DNA region in apoptotic cells. Although the endonuclease responsible for this degradation is not yet identified, a Ca^{++}, Mg^{++}-dependent endonuclease activity can be assayed for in the ventral prostate, and this activity increases substantially during involution (Kyprianou et al. 1988). Since both DNA fragmentation and the Ca^{++}, Mg^{++}-dependent DNase activity increase rapidly in the first 24 hours after castration, whereas apoptotic body accumulation increases markedly after 48 hours, it is likely that specific DNA fragmentation is an early step in prostatic cell apoptosis (English et al. 1989).

Similar to glucocorticoid-mediated apoptosis in lymphocytes (Kaiser and Edelman 1978; McConkey et al. 1989), calcium ion uptake appears to be involved in prostate cell apoptosis, as evidenced by the ability of calcium channel antagonists to suppress ventral prostate involution (Conner et al. 1988; Kyprianou et al. 1988). In accord with the concept that apoptotic cell death is dependent on synthesis of new proteins, the involution of the ventral prostate gland can be suppressed by RNA and protein-synthesis inhibitors (Stanisic et al. 1978; Lee 1981). Therefore, these observations imply that some prod-

FIGURE 1 Photomicrographs of a thin section through the mature rat ventral prostate gland (magnification: top left, 100x; top right, 200x) demonstrate the secretory nature of this exocrine organ in which the luminal ducts are lined with androgen-dependent columnar epithelial cells. 3 days following castration (magnification: lower left, 100x; lower right, 200x), the epithelial cells are in various stages of apoptosis, and the luminal ducts are lined with shrunken epithelial cells and numerous apoptotic bodies (indicated by arrows).

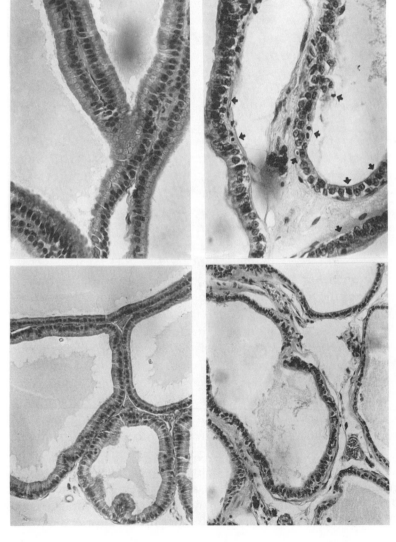

FIGURE 1 (See facing page for legend.)

uct(s) made by the prostate cell after castration is (are) necessary for the onset of the apoptotic process.

Gene Products and Prostate Cell Death

Because such a large proportion of cells in the ventral prostate gland undergo apoptosis following castration, this tissue has been distinctly useful in the search for mammalian gene products specific for apoptosis. Comparative studies using one-

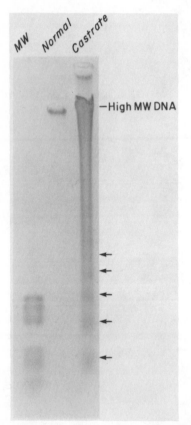

FIGURE 2 DNA extracted from mature (Normal) ventral prostate tissue demonstrates a high-molecular-weight band following electrophoresis in a 1.5% agarose gel, whereas DNA extracted from ventral prostate glands of 3-day castrated rats (Castrate) shows a distinct "ladder" pattern consisting of DNA fragments of sizes in multiples of ~180 bp. (MW) Comparative molecular weight markers.

and two-dimensional SDS-polyacrylamide gel electrophoresis (PAGE) have allowed for the description of several protein spots and bands that appear only during the acute phase of ventral prostate involution (castration-induced proteins) (Anderson et al. 1983; Lee et al. 1985). The identification of proteins from bands or spots following PAGE is, at best, a difficult task, so that many of the castration-induced proteins remain uncharacterized. Certain enzymatic activities, including plasminogen activator (Rennie et al. 1984), acid ribonuclease (Engel et al. 1980), and cathepsin D (Tanabe et al. 1982; Sensibar et al. 1990), have been found to increase during prostatic involution. Each of these specific enzymes is assumed to play a role in the degradative processes associated with prostatic restructuring following castration.

Studies based on isolation and identification of castration-induced mRNAs have proven to be more informative. Two different cDNA species of the involuting prostate have been cloned and sequenced because of their drastically enhanced rate of steady-state synthesis in the acute period following castration. One of these cDNAs encodes a protein of 29 kD identical to the Yb1 subunit of glutathione S-transferase (Chang et al. 1987). Aside from the initial description of this gene product and its pattern of induced expression during prostatic involution, little is known of its cellular site of synthesis in the involuting prostate or its potential role in prostatic apoptosis. On the other hand, a cDNA initially designated "testosterone-repressed prostate message-2" (TRPM-2) (Montpetit et al. 1987) has received increased attention as a marker of apoptosis. In situ hybridization analysis of involuting ventral prostate tissue demonstrates that the expression of this mRNA is limited to the epithelial cells of the involuting gland both prior to and during recognizable changes associated with apoptosis (Buttyan et al. 1989).

If TRPM-2 gene expression is critical to the process of apoptosis, then it would be likely that the product of this gene functions in the metabolic cascade leading to DNA fragmentation and eventual cell death. Sequence studies of TRPM-2 cDNA (Betuzzi et al. 1989; Buttyan et al. 1989; Bandyk et al. 1990a) have revealed that the gene product is identical to a constitutive product of the mammalian Sertoli cell, sulfated glycoprotein-2 (SGP-2) (Collard and Griswold 1987). On the basis of this sequence identity, SGP-2 has been shown to represent one of the castration-induced proteins previously detected by comparative two-dimensional PAGE (Sensibar et al. 1990). This protein has subsequently been found to have many sites of constitutive synthesis in the rat, and it is consistently induced in a variety of cells undergoing apoptosis, regardless of the stimuli used to initiate the apoptotic process. Therefore, it will be discussed as a model gene in the search for the cellular genetic regulatory mechanism that governs the onset of apoptosis.

SGP-2 Gene: A Model for Understanding Apoptotic Regulation of Gene Expression

Aside from the involuting prostate gland, the 2-kb mRNA transcript for SGP-2 (TRPM-2) was also found to be intensely induced in several human and rodent tumors regressing subsequent either to appropriate hormonal treatment or to cytotoxic chemotherapy (Rennie et al. 1988; Buttyan et al. 1989; Kyprianou et al. 1990). In addition, apoptotic cells of the interdigital region in developing fetal rat limb buds express this gene product (Buttyan et al. 1989). Studies of various forms of renal injury have proven most interesting by showing that SGP-2 can be utilized to discriminate cell-specific pathology in a given tissue. Chronic obstruction of a rat ureter leads to the development of hydronephrosis. The hydronephrotic rat kidneys show the highly induced expression of SGP-2 in the epithelial cells of the collecting ducts and distal tubules (Buttyan et al. 1989; Sawczuk et al. 1989), consistent with the described site of apoptosis and cellular deletion (Gobé and Axelson 1987) for this condition. Chronic aminoglycoside administration is toxic to the kidney and activates the expression of SGP-2 in the epithelial cells of the proximal tubules (Bandyk et al. 1990b) corresponding with the development of acute tubular necrosis. Moreover, following experimental renal ischemia/reperfusion injury, SGP-2 expression marks both the proximal and distal tubules as well as other renal cell compartments (Connor et al. 1989).

Perhaps the most intriguing association of SGP-2 expression with tissue involution, based on the independent nature of the discovery, is the report that SGP-2 mRNA is greatly enhanced in human brain tissue affected by Alzheimer's disorder and in hamster brain tissue following scrapie virus infection (Duguid et al. 1989). These disorders have yet to be associated with cellular apoptosis, but the enhanced expression of SGP-2 makes this mechanism a probable candidate for the process of cellular loss leading to these conditions.

As mentioned, SGP-2 is also expressed constitutively in several rodent tissues. Although it is most highly detected in testis and epididymis, other tissues, including liver and brain, show evidence of SGP-2 mRNA and protein expression (Bettuzzi et al. 1989; Grima et al. 1990). A protein sharing immunoidentity with SGP-2 is present in normal rat and human serum (Cheng et al. 1988a; Jenne and Tschopp 1989), and it is not clear to what extent serum perfusion contributes to SGP-2 protein levels detected in many tissues. These multiple sites of expression have resulted in a confusing array of terms applied to this product by different investigators studying different organs and tissues. Aside from the TRPM-2 designation, which is based on its isolation from involuting prostate tissue, SGP-2 is also referred to under the names of SP40-40 (Jenne and Tschopp 1989; Kirszbaum et al. 1989), clusterin (Cheng et al. 1988b), and apolipoprotein J (de Silva et al. 1990). Comparative analysis of the polypeptide encoded by this gene shows that it is a member of a protein family that includes human apolipoprotein A1 (Collard and Griswold 1987). In addition, SGP-2 has limited homology with complement proteins, C7, C8, and C9 (Kirszbaum et al. 1989). As for as its actual role in apoptosis, the abundant synthesis of SGP-2 by normal Sertoli cells implies that this protein is not likely to be a potentially lethal product as has been hypothesized to participate in the onset of apoptosis. In fact, recent studies on the function of the SGP-2 protein give reason to believe that this substance might be synthesized to protect surrounding areas of remodeling tissues.

Functional studies of SGP-2 protein from serum and the reproductive tract show that it is a potent inhibitor of complement-mediated cytolysis (Jenne and Tschopp 1989; Kirszbaum et al. 1989). This activity has led to speculation that the binding of SGP-2 to the sperm membrane prior to ejaculation would help to protect sperm from complement attack in the female reproductive tract (Jenne and Tschopp 1989; Kirszbaum et al. 1989). Therefore, the function of SGP-2 in the male reproductive tract might be related to its ability to block complement activity against sperm during fertilization.

A need for this same function may be why SGP-2 is expressed so ubiquitously during apoptosis. Nonspecific complement activation has proven to be one of the more damaging components of the ischemic process in cardiac tissue (Crawford et al. 1988). By infusing a potent complement inhibitor prior to experimental myocardial ischemia, one can temper the subsequent injury to the heart (Weisman et al. 1990). Therefore, any speculation on the reason that SGP-2 is so intensely expressed in remodeling tissues should include the suggestion that this protein is synthesized by the dying cells in an attempt to protect surrounding tissues from nonspecific complement activation such as might occur when basement membranes are exposed in involuting tissues.

Regardless of its actual role in apoptosis, SGP-2 continues to show a strong association with this process. Since it is not inconsistent that a gene product might be expressed under two drastically different conditions (for example, mammalian heat shock genes that are induced by stressful conditions and expressed constitutively during mammalian gametogenesis), SGP-2 should provide a valuable model to identify the genetic regulatory elements governing apoptosis-specific expression of gene products in the future. As additional genes expressed during apoptosis continue to be identified (transglutaminase, glutathione S-transferase, and others), a comparison of the genomic regulatory regions of these genes might lead to the identification of consensus sequences for apoptosis-specific expression.

Growth Factor and Proto-oncogene Expression during Prostatic Involution

Another approach to the study of genetic activity associated with apoptosis has involved the search for the synthesis of prespecified gene products during prostatic involution. Comparative Northern blot analysis of involuting ventral prostate mRNA has been used to demonstrate that RNA transcripts encoding transglutaminase (L. Fesus, pers. comm.) and transforming growth factor-β (TGF-β) (Kyprianou and Isaacs 1989) are transiently induced in this tissue, corresponding with the peak period of apoptosis. Both these substances have been previously associated with apoptotic cells or areas of cellular injury. Transglutaminase expression is thought to prepare apoptotic cells in the liver for phagic digestion (Fesus et al. 1989), and TGF-β synthesis during myocardial ischemia has been proposed to provide cardioprotection by blocking tumor necrosis factor action (Lefer et al. 1990).

Aside from the demonstration of enhanced TGF-β expression in the involuting ventral prostate, there is little understanding as to how castration might alter the milieu of other growth factors available to the prostatic cells. Two types of growth factors can be readily identified and assayed for in adult rat ventral prostate tissue: basic fibroblast growth factor and an epidermal growth-factor-like substance (Jacobs et al. 1988). If we assume that testosterone action on the prostate is mediated by its ability to regulate local growth factor synthesis, it is then possible that prostatic apoptosis is initiated by aberrations in the availability of these growth factors. Clearly, this is an area in which more research is required.

Several proto-oncogenes are transiently expressed during the early period of prostatic involution, as shown by both Northern blot and in situ hybridization studies. In a pattern that reiterates the genetic activity of ventral prostate cells stimulated to proliferate by androgen supplementation (Katz et al. 1989), androgen-deprived ventral prostate cells show induced expression of c-*fos*, c-*myc*, and heat shock 70K transcripts in a cascade-like temporal order (Buttyan et al. 1988). In situ hybridization analysis has confirmed the postcastration-induced expression of c-*myc* transcripts and identified the luminal epithelial cells as the source of this activity in 3-day involuting ventral prostate (Quarmby et al. 1987).

In a sense, these results present a paradox; this cascade pattern of gene activity has often been associated with cells responding to a proliferative stimulus. Therefore, what relevance could this gene pattern have to cells about to undergo apoptosis? One possible explanation comes from the unique situation of having essentially noncycling cells (the mature rat ventral prostate gland has a very low rate of cellular proliferation) seemingly capable of directly entering the apoptotic pathway. This phenomenon is in contradistinction to other sites of apoptosis, most prominently chemotherapeutic-induced apoptosis, wherein cells must be in a proliferative cycle to undergo this process (Eastman 1990; Kung et al. 1990). Perhaps the first step of the differentiated prostate cell onto the pathway of apoptosis is a commitment to reenter the cell cycle from the hypothetical G_0 state. Therefore, the pattern of proto-oncogene expression in ventral prostate epithelial cells early after castration may be a manifestation of commitment through the early cell cycle. If this is the case, the apoptotic prostate cell must deviate from the normal proliferative pathway at some point prior to S phase, since little thymidine incorporation can be detected in the ventral prostate gland during the acute period following castration (Fig. 3). The concept of apoptosis resulting from the inability to complete the normal cell cycle has been previously considered as an explanation of glucocorticoid-induced apoptosis of lymphocytes (Harmon et al. 1979).

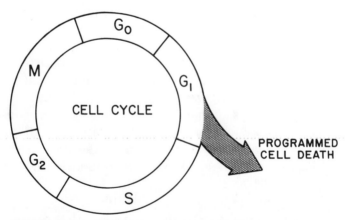

FIGURE 3 Graphic representation of the concept of prostatic apoptosis as a deviation from the normal proliferative cell cycle.

CONCLUSION

To date, in vivo studies of the involuting ventral prostate gland have provided a foundation for further studies on the molecular basis of cellular apoptosis. The large number of prostatic cells that become apoptotic after androgen withdrawal, in addition to the predictable period in which apoptosis occurs, has enabled the isolation and identification of apoptosis-specific gene products and a cursory description of physical and molecular behavior corresponding to the early period of commitment to apoptosis. Although the signal (androgen deprivation) that initiates the apoptotic response in prostate cells is very different than for other systems of apoptosis, shared patterns of physiological changes, biochemical markers, and genetic activity between apoptotic prostate cells and other types of cells undergoing apoptosis suggest some commonality in the molecular mechanism of the apoptotic pathway. Identification and cloning of gene products that are induced specifically during the apoptotic process should enable the characterization of apoptosis-specific regulatory elements of DNA. Most of all, the use of the regressing ventral prostate to study the apoptotic response is leading us to the understanding that this process must mimic the proliferative cell cycle, at least in terms of its complexity.

REFERENCES

Anderson, K.M., J. Baranowski, S.G. Economous, and M. Rubenstein. 1983. A qualitative analysis of acidic proteins associated with regressing, growing or dividing rat ventral prostate cells. *Prostate* **4**: 151.

Bandyk, M.G., I.S. Sawczuk, C.A. Olsson, A.E. Katz, and R. Buttyan. 1990a. Characterization of the products of a gene expressed during androgen-programmed cell death and their potential use as a marker of urogenital injury. *J. Urol.* **143**: 407.

Bandyk, M., R. Buttyan, C.A. Olsson, G. Appel, V. D'Agati, A. Katz, P.-Y. Ng, and I.S. Sawczuk. 1990b. TRPM-2 detection and localization during gentamicin-induced nephrotoxicity. *J. Urol.* **143**: 239A.

Bettuzzi, S., R.A. Hiipakka, P. Gilna, and S. Liao. 1989. Identification of an androgen-repressed mRNA in rat ventral prostate as coding for sulfated glycoprotein 2 by cDNA cloning and sequence analysis. *Biochem. J.* **257**: 293.

Bruchovsky, N., B. Lesser, and E. van Doorne. 1975. Hormonal effects on cell proliferation in rat prostate. *Vitam. Horm.* **33:** 61.

Buttyan, R., Z. Zakeri, R. Lockshin, and D. Wolgemuth. 1988. Cascade induction of c-*fos*, c-*myc* and heat shock 70K transcripts during regression of the rat ventral prostate gland. *Mol. Endocrinol.* **2:** 650.

Buttyan, R., C.A. Olsson, J. Pintar, C. Chang, M. Bandyk, P.-Y. Ng, and I.S. Sawczuk. 1989. Induction of the *TRPM-2* gene in cells undergoing programmed cell death. *Mol. Cell. Biol.* **9:** 3473.

Crawford, M.H., F.L. Glover, W.P. Kolb, A. McMahon, R.A. O'Rourke, L.M. McManus, and R.N. Pinkard. 1988. Complement and neutrophil activation in the pathogenesis of ischemic myocardial injury. *Circulation* **78:** 1449.

Chang, C., A.G. Saltzman, N.S. Sorensen, R.A. Hiipakka, and S. Liao. 1987. Identification of glutathione S-transferase Yb_1 mRNA as the androgen-repressed mRNA by cDNA cloning and sequence analysis. *J. Biol. Chem.* **262:** 11901.

Cheng, C.Y., P.P. Mathur, and J. Grima. 1988a. Structural analysis of clusterin and its subunits in ram rete testis fluid. *Biochemistry* **27:** 4079.

Cheng, C.Y., C.-L.C. Chen, Z.-M. Feng, A. Marshall, and C.W. Bardin. 1988b. Rat clusterin isolated from primary sertoli cell-enriched culture medium is sulfated glycoprotein-2 (SGP-2). *Biochem. Biophys. Res. Commun.* **155:** 398.

Coffey, D.S., J. Shimazaki, and H.G. Williams-Ashman. 1976. Polymerization of deoxyribonucleotides in relation to androgen-induced prostatic growth. *Arch. Biochem. Biophys.* **308:** 426.

Collard, M.W. and M.D. Griswold. 1987. Biosynthesis and molecular cloning of sulfated glycoprotein 2 secreted by rat sertoli cells. *Biochemistry* **26:** 3297.

Connor, J., R. Buttyan, C.A. Olsson, K. O'Toole, P.-Y. Ng, and I.S. Sawczuk. 1989. Characterization and modulation of renal gene activity in response to acute renal ischemia. *Surg. Forum* **40:** 684.

Connor, J., I.S. Sawczuk, M.C. Benson, P. Tomashefsky, K.M. O'Toole, C.A. Olsson, and R. Buttyan. 1988. Calcium channel antagonists delay regression of androgen-dependent tissues and suppress gene activity associated with cell death. *Prostate* **13:** 119.

de Silva, H.V., J.A.K. Harmony, W.D. Stuart, C.M. Gil, and J. Robbins. 1990. Apolipoprotein J: Structure and tissue distribution. *Biochemistry* **29:** 5380.

Duguid, J.R., C.W. Bohmont, N. Liu, and W.W. Tourtellotte. 1989. Changes in brain gene expression shared by scrapie and Alzheimer disease. *Proc. Natl. Acad. Sci.* **86:** 7260.

Eastman, A. 1990. Activation of programmed cell death by anticancer agents: Cisplatin as a model system. *Cancer Cells* **2:** 275.

Engel, G., C. Lee, and J.T. Grayhack. 1980. Acid ribonuclease in rat prostate during castration-induced involution. *Biol. Reprod.* **22:** 827.

English, H.F., J.R. Drago, and R.J. Santen. 1985. Cellular response to androgen depletion and repletion in the rat ventral prostate: Autoradiography and morphometric analysis. *Prostate* **7:** 41.

English, H.F., N. Kyprianou, and J.T. Isaacs. 1989. Relationship between DNA fragmentation and apoptosis in the programmed cell death in the rat prostate following castration. *Prostate* **15:** 233.

Fesus, L., V. Thomazy, F. Autuori, M.P. Ceru, E. Tarcsa, and M. Piacentina. 1989. Apoptotic hepatocytes become insoluble in detergents and chaotropic agents as a result of transglutaminase action. *FEBS Lett.* **245:** 150.

Gobé, G.C. and R.A. Axelsen. 1987. Genesis of renal tubular atrophy in experimental hydronephrosis in the rat. Role of apoptosis. *Lab. Invest.* **56:** 273.

Grima, J., I. Zwain, R.A. Lockshin, C.W. Bardin, and C.Y. Cheng. 1990. Diverse secretory patterns of clusterin by epididymis and prostate/seminal vesicles undergoing regression after orchiectomy. *Endocrinology* **126:** 2989.

Harmon, J.M., M.R. Norman, B.J. Fowlkes, and E.B. Thompson. 1979. Dexamethasone induces irreversible G1 arrest and death of a human lymphoid cell line. *J. Cell. Physiol.* **98:** 267.

Helminen, H.J. and J.L.E. Ericcson. 1972. Ultrastructural studies on prostatic involution in the rat. Evidence for focal irreversible damage to epithelium and heterophagic digestion in macrophages. *J. Ultrastruct. Res.* **39:** 443.

Huggins, C. and C.V. Hodges. 1941. Studies on prostatic cancer. I. The effect of castration, of estrogen and of androgen injection on serum phosphatases in metastatic carcinoma of the prostate. *Cancer Res.* **1:** 293.

Isaacs, J.T. 1984. Antagonistic effect of androgen on prostatic cell death. *Prostate* **5:** 545.

Jacobs, S.C., M.T. Story, M. Sasse, and R. Lawson. 1988. Characterization of growth factors derived from the rat ventral prostate. *J. Urol.* **139:** 1106.

Jenne, D.E. and J. Tschopp. 1989. Molecular structure and functional characterization of a human complement cytolysis inhibitor found in blood and seminal plasma: Identity to sulfated glycoprotein 2, a constituent of rat testis fluid. *Proc. Natl. Acad. Sci.* **86:** 7123.

Kaiser, N. and I.S. Edelman. 1978. Further studies on the role of calcium in glucocorticoid-induced lymphocytolysis. *Endocrinology* **103:** 936.

Katz, A.E., M.C. Benson, G.J. Wise, C.A. Olsson, M.G. Bandyk, I.S. Sawczuk, P. Tomashefsky, and R. Buttyan. 1989. Gene activity during the early phase of androgen-stimulated rat prostate regrowth. *Cancer Res.* **49:** 5889.

Kerr, J.F.R. and J. Searle. 1973. Deletion of cells by apoptosis during castration-induced involution of the rat prostate. *Virchows Arch. Zellpathol.* **13:** 87.

Kiplesund, K.M., J. Halqunset, H.R. Fjosne, and A. Sunde. 1988. Light microscopic morphometric analysis of castration effects in the different lobes of the rat prostate. *Prostate* **13:** 221.

Kirszbaum, L., J.A. Sharpe, B. Murphy, A.J.F. d'Apice, B. Classon, P. Hudson, and I.D. Walker. 1989. Molecular cloning and character-ization of the novel human complement-associated protein, SP40,40: A link between the complement and the reproductive systems. *EMBO J.* **8:** 711.

Kung, A.L., A. Zetterberg, S.W. Sherwood, and R.T. Schimke. 1990. Cytotoxic effects of cell cycle phase specific agents: Results of cell cycle perturbation. *Cancer Res.* **50:** 7307.

Kyprianou, N. and J.T. Isaacs. 1988. Activation of programmed cell death in the rat ventral prostate after castration. *Endocrinology* **122:** 552.

————. 1989. Expression of transforming growth factor B in the rat ventral prostate during castration-induced programmed cell death. *Mol. Endocrinol* **3:** 1515.

Kyprianou, N., H.F. English, and J.T. Isaacs. 1988. Activation of a Ca^{++}-Mg^{++}-dependent endonuclease as an early event in castration-induced prostate cell death. *Prostate* **13:** 103.

————. 1990. Programmed cell death during the regression of PC-82 human prostate cancer following androgen ablation. *Cancer Res.* **50:** 3748.

Lee, C. 1981. Physiology of castration induced regression in the rat prostate. *Prog. Clin. Biol. Res.* **75:** 145.

Lee, C., Y. Tsai, H.H. Harrison, and J. Sensibar. 1985. Proteins of the rat prostate: I. Preliminary characterization by two-dimensional electrophoresis. *Prostate* **7:** 171.

Lefer, A.M., P. Tsao, N. Aoki, and M.A. Palladino, Jr. 1990. Mediation of cardioprotection by transforming growth factor β. *Science* **249:** 61.

Lesser, B. and N. Bruchovsky. 1973. The effects of testosterone, 5-a-dihydrotestosterone and adenosine 3´, 5´-monophosphate on cell proliferation and differentiation in rat prostate. *Biochem. Biophys. Acta* **308:** 426.

McConkey, D.J., P. Hartzell, M. Jondal, and S. Orrenius. 1989. In-hibition of DNA fragmentation in thymocytes and isolated thymocyte nuclei by agents that stimulate protein kinase C. *J. Biol. Chem.* **264:** 13399.

Montpetit, M.L., K.R. Lawless, and M.P. Tenniswood. 1987. Androgen-repressed messages in the rat ventral prostate. *Prostate* **8:** 25.

Parker, M.G. and G.T. Scrace. 1979. Regulation of protein synthesis in rat ventral prostate: Cell-free translation of mRNA. *Proc. Natl.*

Acad. Sci. **76:** 1580.

Quarmby, B.E., W.C. Beckman, E.M. Wilson, and F.S. French. 1987. Androgen regulation of *c-myc* messenger ribonucleic acid levels in rat ventral prostate. *Mol. Endocrinol.* **1:** 865.

Rennie, P.S., R. Bouffard, N. Bruchovsky, and H. Cheng. 1984. Increased activity of plasminogen activators during involution of the rat ventral prostate. *Biochem. J.* **221:** 171.

Rennie, P.S., N. Bruchovsky, R. Buttyan, M. Benson, and H. Cheng. 1988. Gene expression during the early phases of regression of the androgen-dependent Shionogi mouse mammary carcinoma. *Cancer Res.* **48:** 6309.

Sanford, M.L., J.E. Searle, and J.F.R. Kerr. 1984. Successive waves of apoptosis in the rat prostate after repeated withdrawal of testosterone stimulation. *Pathology* **16:** 406.

Sawczuk, I.S., G. Hoke, C.A. Olsson, J. Connor, and R. Buttyan. 1989. Gene expression in response to acute unilateral ureteral obstruction. *Kidney Int.* **35:** 1315.

Sensibar, J.A., X. Liu, B. Patai, B. Alger, and C. Lee. 1990. Characterization of castration-induced cell death in the rat prostate by immunohistochemical localization of cathepsin D. *Prostate* **16:** 263.

Stanisic, T.R., R. Sadlowsky, C. Lee, and J.T. Grayhack. 1978. Partial inhibition of castration induced ventral prostate regression with actinomycin D and cycloheximide. *Invest. Urol.* **16:** 19.

Sufrin, G. and D.S. Coffey. 1973. A new model for studying the effects of drugs on prostatic growth. I. Antiandrogens and DNA synthesis. *Invest. Urol.* **11:** 45.

Tanabe, E., C. Lee, and J.T. Grayhack. 1982. Activities of cathepsin D in rat prostate during castration-induced involution. *J. Urol.* **127:** 826.

Weisman, H.F., T. Bartow, M.K. Leppo, H.C. Marsh, G.R. Carson, M.F. Concino, M.P. Boyle, K.H. Roux, M.C. Weisfeldt, and D.T. Fearon. 1990. Soluble human complement receptor type 1: In vivo inhibitor of complement suppressing post-ischemic myocardial inflammation and necrosis. *Science* **249:** 146.

Apoptosis and Cell Proliferation Are Terms of the Growth Equation

L.E. Gerschenson and R.J. Rotello
Department of Pathology, School of Medicine
University of Colorado Health Sciences Center
Denver, Colorado 80262

Our laboratory has been involved for several years in the study of growth regulation in cells of the rabbit uterine epithelium. The aim was to define the role(s) of estrogens and progesterone in regulating normal cell functions of proliferation, migration, differentiation, and death. The rationale behind these studies has been that derangement of normal cell functions may be a cause of cancer. There is a high incidence of spontaneous endometrial carcinoma in rabbits (Greene 1959), and estrogens appear to play a role in promoting endometrial carcinoma (Baba and von Haam 1967; Siiteri 1978), whereas progesterone has a clear inhibitory effect on that disease (Baba and von Haam 1967; Greenblatt et al. 1982). These hormonal effects have been observed in both humans and rabbits. In this paper, we review our own work and discuss our thoughts on the meaning of apoptosis.

Cell Proliferation

Several years ago, we described an experimental model of primary cultured rabbit uterine epithelial cells using serum-free media (Gerschenson et al. 1974). Estrogen addition was found to have a discrete proliferative effect on quiescent cells, and progesterone was found to antagonize that effect (Gerschenson et al. 1976, 1977). This phenomenon appeared to be mediated by a factor induced by the action of progesterone on a dividing cell population. In order to account for this apparent antagonism, we proposed that an "estrogen inhibitory factor" was induced by progesterone (Gerschenson et al. 1979), which then interfered with cell proliferation (Fig. 1).

Apoptosis: The Molecular Basis of Cell Death
Copyright 1991 Cold Spring Harbor Laboratory Press 0-87969-366-5/91 $3.00 + 00

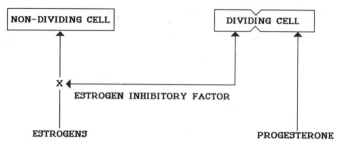

FIGURE 1 Progesterone elicits the production of an estrogen inhibitory factor on dividing cells. This factor inhibits the proliferative effect of estrogens on nondividing or quiescent cells.

In subconfluent cultures, which were plated at higher densities, estrogen did not elicit a proliferative response. This inhibition of the proliferative response was specific for estrogens, since prostaglandin $F_{2\alpha}$ and epidermal growth factor were each found to increase cell proliferation in both low- and high-density cell cultures. Inhibitory activity was detected in conditioned media as well as in cell homogenates from high- but not low-density cultures. This activity was found to be heat- and trypsin-sensitive, nondialyzable, and precipitable at high concentrations of ammonium sulfate, and it did not appear to be mediated directly through estrogen receptors, since no decrease in estrogen binding was detected under the influence of the inhibitor (Fig. 2) (Gerschenson et al. 1981, 1984).

The experiments described above supported the concept that factors were produced which could modulate the proliferative response to estrogens. It is postulated that these estrogen inhibitory factors may in fact be "cell death factors" (see below), that is, these factors may induce increased cell death, perhaps compensating for the proliferation induced by estrogens.

Experiments were conducted with rabbits to see if our "in culture" findings could be extrapolated to the "in vivo" situation. As expected, these studies demonstrated that both estrogen and progesterone regulate the proliferation of cells in the rabbit uterine epithelium. Whereas estradiol-17β stimulated the proliferation of the endometrium gland cells, progesterone had a general effect of enhancing proliferation in both glandular and luminal epithelial cells. However, simulta-

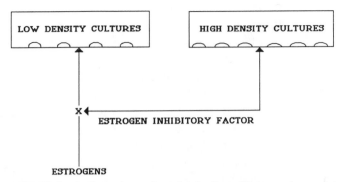

FIGURE 2 Cells cultured at high densities produce an estrogen inhibitory factor, which inhibits the proliferative effect of estrogens on cells cultured at low densities.

neous administration of both hormones resulted in mutual antagonism. We not only detected geographical target cell differences in response to the hormones, but we also determined that estradiol-17β exerts its proliferative effect on quiescent cells, recruiting them into the cell cycle, whereas progesterone acts on actively proliferating cells (Gerschenson et al. 1979; Conti et al. 1981, 1984; Murai et al. 1981). Both estradiol and progesterone individually were found to increase cell proliferation; however, the end point of the estrogenic effect is to generate progesterone-responsive cells, whereas progesterone treatment leads to an increase in nondividing, differentiated cells (Murai et al. 1981). Therefore, both steroid hormones regulate rabbit uterine epithelium growth in a coordinated fashion, which results finally in a differentiated and nonproliferating epithelium.

Cell Migration

Our studies showed that endometrial gland epithelial cells migrate toward the luminal surface at a rate that is accelerated by estrogen administration. This was demonstrated by administration of a single dose of [³H]thymidine to rabbits, followed by determination at selected times of the geographical distribution of cells with labeled nuclei. As a function of time, there was a decrease in the number of labeled cells in the bottom of the glands with a concomitant increase in the upper part of the glandular and luminal epithelium (Conti et al. 1984). These observations are in accord with the view that progesterone administration resulted in the proliferation of luminal cells, which then formed new glands that penetrated (or invaded) the stroma until they stopped proliferating and became fully differentiated, as shown by the synthesis and secretion of uteroglobin (Murai et al. 1981). In control animals, cells migrate from the bottom of "stem glands" to the lumen, whereas their progeny invade the stroma during the progesterone-dependent secretory development of the rabbit endometrium. Therefore, cell migration is a normal component of rabbit epithelium growth, and it is regulated by ovarian steroid hormones.

Cell Differentiation

The uterus undergoes marked morphological and physiological changes under the influence of progesterone. These changes are thought to be associated with preparation for blastocyst implantation. In the early pregnant or pseudopregnant rabbit, the protein pattern in the uterine secretions is dominated by a 15,000 m.w. glycoprotein of unknown function called uteroglobin (Beier 1968). Uteroglobin has provided a specific differentiation marker for the study of uterine epithelium biology and has proven to be a valuable tool in that respect.

Using primary cultures of rabbit endometrium, as well as animal studies, we demonstrated that uteroglobin is synthesized by nondividing glandular epithelial cells (Murai et al. 1981; Rajkumar et al. 1983a,b). This increase in uteroglobin production appears to be achieved both by increased synthesis of uteroglobin and by an increased number of cells involved in that synthesis. These phenomena are regulated by progesterone (Shroyer et al. 1987b). This hormone not only regulates uteroglobin production at a transcriptional level, but also regulates its mode of secretion. Estradiol-17β administration moderately increases the synthesis of uteroglobin; however, its secretion mode is different from that induced by progesterone (Shroyer et al. 1987a). At different times after inducing increased production of uteroglobin with progesterone, cells intensely immunostained for uteroglobin were observed. Ultra-structural studies, the failure to exclude trypan blue, and the presence of free cytoplasmic uteroglobin revealed that these cells into which uteroglobin freely penetrated (Shroyer et al. 1987b) were either dead or dying. As discussed below (see Fig. 3), we believe that the differentiated cells are the target for the apoptotic stimuli.

Cell Death

In 1984, we published data showing that estradiol-17β administration diminished the rate of cell loss in the luminal epithelium of the rabbit endometrium (Conti et al. 1984). At that time, it was not evident that cell death regulation could play a role in uterine cell growth regulation by hormones. While studying the biology of cultured primary uterine epithelial cells grown on floating collagen gels in serum-free media, we observed a novel growth behavior. The cultures exhibited cyclic changes in DNA content throughout long periods of cell culture (35 days). These cycles were characterized by significant increases and decreases in cellular DNA content or in the number of cells present; however, throughout the duration of the culture, there was no net change in the total amount of DNA present. The dead cells were observed through the use of light microscopy and appeared to be apoptotic on the basis of their morphology. The rates of cell death and proliferation changed with respect to time in culture with the same periodicity as the DNA changes. Thus, to maintain homeostasis in the total culture mass, a feedback control mechanism can be postulated, presumably mediated by soluble cell proliferation factor(s) and cell death factor(s) produced by the cells (Lynch et al. 1986). These factors have been detected both in conditioned media and in extracts from cultured cells. Both activities are heat- and trypsin-sensitive. These experimental data suggest that primary endometrial cell cultures are made up of a renewing cell population containing stem cells and their progeny (Gerschenson et al. 1979; Conti et al. 1984). The growth of these cells is modulated by both cell proliferation and death factors, which act in a paracrine fashion (Lynch et al. 1986).

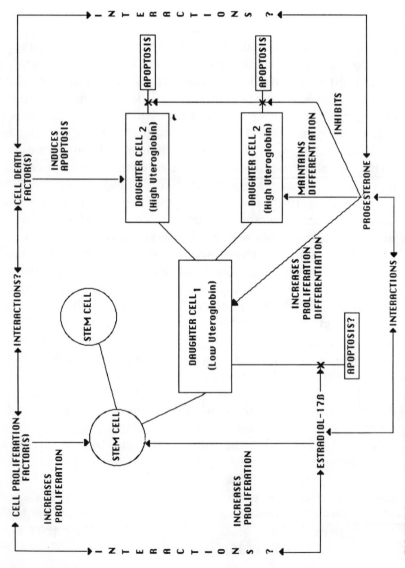

FIGURE 3 (See facing page for legend.)

The above data stimulated further in vivo cell death research. In these studies, it was determined that administration of estradiol-17β or progesterone to normal female rabbits
resulted not only in increased cell proliferation of rabbit
uterine epithelium, but also in decreased number of apoptotic
cells. However, progesterone was far more efficient than
estradiol-17β in suppressing apoptosis (Nawaz et al. 1987).

A single dose of human chorionic gonadotropin (hCG) given
to female rabbits induces pseudopregnancy, which entails
ovulation, formation of the corpus luteum, and increased
serum levels of progesterone. Four days after hCG administration, the uterine epithelium is nonproliferating and fully differentiated, showing arborization and secretory characteristics
(Murai et al. 1981; Conti et al. 1984). Ovariectomy at that time
was found to induce about a 100-fold increase in the number
of apoptotic uterine epithelial cells. Progesterone administration prevented such an increase (Nawaz et al. 1987; Rotello et
al. 1989). It should be noted that in rabbit uterine epithelium,
apoptosis is the predominant type of cell death (97.5%), as opposed to necrosis (2.5%) (Nawaz et al. 1987; Rotello et al.
1989).

Since morphological studies could be considered subjective,
we sought to determine the presence of the specific apoptotic
DNA degradation pattern (Rotello et al. 1989). DNA isolated
from endometrium of ovariectomized pseudopregnant rabbits
showed a pattern of DNA cleavage at internucleosomal locations. DNA from the endometrium of nonovariectomized animals, as well as from several other organs, did not exhibit
such a pattern. Only the uterine epithelium showed significant
apoptotic cell death, which is virtually absent in the stromal
compartment (Rotello et al. 1989).

As early as 1 day after ovariectomy of pseudopregnant rabbits, apoptotic cells and apoptotic bodies are seen in the
uterine epithelium. These characteristic cells with condensed

FIGURE 3 Diagram of demonstrated and hypothetical regulation of
uterine epithelial cell proliferation, differentiation, and death by ovarian steroid hormones and paracrine factors. Only two proliferation
cycles are shown for brevity, but more cycles occur normally.

and fragmented nuclei are seen among normal intact epithelial cells and without any inflammatory cell infiltrate. Agarose gel electrophoresis of DNA isolated from endometrial scrapings revealed that the characteristic oligonucleosomal DNA fragments could be detected as early as 1 day postovariectomy. At this time, approximately 40% of the isolated DNA is fragmented, as measured by densitometric scanning of the negative images obtained from photographing ethidium-bromide-stained agarose gels. By 1.5 days after ovariectomy, approximately 15% of the uterine epithelial cells were apoptotic by morphological criteria, whereas almost 70% of the isolated DNA was observed to be fragmented. Maximal levels of DNA fragmentation were detected at 1.5 days postovariectomy, whereas maximal numbers of apoptotic cells were observed at 2 days postovariectomy. At 4 days postovariectomy, DNA fragmentation was undetectable, and the number of apoptotic cells was similar to that of control levels, which is consistently 0.2–0.5% of total uterine epithelial cells. Finally, a simple uterine epithelium remained at 12 days postovariectomy, and the uterine tissue mass had significantly regressed to approximately one-third of control pseudopregnant uterine mass. Apoptotic cells observed at selected times after ovariectomy were either taken up by neighboring epithelial cells or sloughed off into the uterine lumen. We have found that virtually no apoptotic cells are phagocytosed by macrophages, which do not appear to migrate significantly to the uterus to aid in the removal of apoptotic cells or bodies (Nawaz et al. 1987; Rotello et al. 1991a).

In most species, progesterone is the important hormone involved in maintaining breast and endometrial growth during pregnancy. In addition to its role in pregnancy, a connection between apoptosis and progesterone is implicated in the observation that the number of apoptotic cells increases in human premenstrual human endometrium (Hopwood and Levison 1976). On the basis of this observation and our studies with rabbits, it appears that progesterone promotes cell proliferation and differentiation and suppresses apoptosis by an unknown mechanism. In contrast to the situation in control ovariectomized animals, when progesterone was administered for 2 days to an ovariectomized animal, neither apoptotic cells nor DNA fragmentation was detectable. Various other steroid hormones administered alone did not suppress apoptosis, including estradiol-17β, hydrocortisone, testosterone, tamoxifen, and 20α-OH progesterone, a major progestin metabolite of rabbit ovaries (Rotello et al. 1991a). Further evidence for progesterone regulation of apoptosis is provided by studies with the synthetic progesterone antagonist, RU 486. Its administration to pseudopregnant animals for 2 days induced apoptosis both morphologically (25% apoptotic cells vs. about 0.2% in control animals) and biochemically (61% DNA fragmentation vs. no detectable fragmentation in control animals). These values are similar to those observed in pseudopregnant ovariectomized animals. RU 486 binds the progesterone receptor with high affinity, and the resultant complex is capable of binding DNA (Guiochon-Mantel et al. 1988). However, RU 486-progesterone receptor complexes have not been found to have similar binding activity in intact cells (Baulieu 1989), a fact that may explain the inhibition of progesterone action.

Endonuclease activity. At the selected times after ovariectomy, there is increased endonuclease activity above control levels in nuclear protein extracts from endometrial scrapings. This activity has been observed by analyzing the nicking of a supercoiled plasmid DNA and subsequent agarose gel electrophoresis, or by measuring the release of acid-soluble plasmid [^{3}H]DNA (Low et al. 1988). In each of the aforementioned assays, endonuclease activity appears to be more Mg^{++}- than Ca^{++}-dependent (Rotello et al. 1990). Similar data have been obtained recently using permeabilized nuclei from a rabbit endometrial cell line as a substrate (R.J. Rotello, unpubl.). In other systems described to date, both Ca^{++} and Mg^{++} appear to be required for activation of an endogenous endonuclease in isolated nuclei (Hewish and Burgoyne 1973; Wyllie 1980); however, a reduced cationic environment in isolated nuclei can result in DNA fragmentation (Vanderbilt et al. 1982). Arends et al. (1990) have recently described results where alterations in chromatin appear not to be responsible for an endonuclease's gaining access to DNA, but instead, DNA fragmentation may be the result of an endonuclease's being activated and cleaving DNA into oligonucleosomal fragments.

Apoptosis in cell cultures. Uterine epithelial cells can be established as primary cultures on plastic tissue culture dishes or on floating collagen gels. Cells cultured on collagen gels were the first culture condition in which apoptosis morphology was observed. In addition, conditioned media experiments suggested that soluble factors could increase either proliferation or apoptosis, depending on the time of culture (see above).

Recently, in primary uterine cells cultured on plastic, apoptosis was identified biochemically by the presence of the characteristic patterns of DNA fragmentation. The levels of DNA fragmentation in culture are measurable by scanning the negative image of a photograph of an ethidium-bromide-stained DNA gel with a video densitometer. Although we have not yet studied a hormonal (i.e., progesterone) regulation of apoptosis in culture, we have found that serum can suppress apoptosis in a concentration-dependent and species-independent manner. Primary uterine cells incubated with 5% calf serum exhibit 0–5% DNA fragmentation versus 45–50% in cells incubated in chemically defined media without serum. This serum effect is not inhibited by addition of RU 486, which suggests that progesterone in serum may not be the only factor suppressing apoptosis in culture. However, incubating serum-treated primary uterine epithelial cells with the RNA synthesis inhibitor actinomycin D, or with protein synthesis inhibitors cycloheximide or puromycin, resulted in a significant induction of apoptosis with 65–70% DNA fragmentation. Various cell lines or isolated permeabilized nuclei incubated with these macromolecular synthesis inhibitors did not induce DNA fragmentation. These studies therefore suggest that RNA and protein synthesis inhibitors interfere with the serum suppression of primary uterine epithelial cell apoptosis, possibly by preventing the synthesis of an inhibitor of the cell death program. It is unlikely that these agents with different mechanisms of action induce apoptosis randomly through interference with other cellular machinery. In contrast to our model, apoptosis that occurs in nerve-growth-factor-deprived neurons can be prevented by the addition of RNA and protein synthesis inhibitors, suggesting a need for new protein synthesis to bring about the death of neurons (Martin et al. 1988). In agreement with our studies, Martin et al. (1990) have shown that HL-60 human leukemic cells that have been induced to differentiate undergo apoptosis when incubated with actinomycin D or cycloheximide, which also suggests that an inhibitor of the death program may now be absent and incapable of suppressing cell death. In addition, in support of the hypothesis that an inhibitor or repressor of the cell death program exists, increased numbers of apoptotic cells have been observed when certain agents that interfere with RNA and protein synthesis are added to respective tissues on cells. For example, increased apoptotic cells were seen in rabbit intestines of animals given the shiga toxin (Keenan et al. 1986). This toxin is believed to interfere with the 60S ribosomal subunit involved in protein synthesis. Another example is that of the plant toxins ricin and abrin, each of which appears to increase the incidence of apoptotic-like cells seen in mammalian lymphoid tissues and intestines (Griffiths et al. 1987). These toxins are potent inhibitors of protein synthesis. Therefore, the aforementioned reports and our data indicate that apoptosis suppression requires RNA and protein synthesis and suggest a negative regulation of apoptosis.

Transforming growth factor β1 and apoptosis in cell cultures. The family of transforming growth factor β (TGF-β) molecules has been strongly implicated in embryonic development and wound healing (Roberts et al. 1988). In our experimental cell culture model, rTGF-β1 induces apoptosis. This action of rTGF-β1 appears to be twofold: inhibition of cell proliferation and induction of apoptosis (Rotello et al. 1991b). The action of rTGF-β1 appears to be concentration-dependent and can be partially neutralized by incubating with antibodies against rTGF-β1. The addition of rTGF-β1 induces levels of cell death that approach what is seen when serum is withdrawn from cultures (55–60% DNA fragmentation). However, the death-inducing capability is highest when rTGF-β1 is added in the presence of 0.05% serum. We also have preliminary evidence that Müllerian inhibitor substance (MIS) can induce apoptosis in primary uterine epithelial cell cultures (Rotello et al. 1990), and this finding suggests that molecules in the TGF-β1 family may have similar functions on target cells.

DISCUSSION

This brief summary of our experimental data shows that our involvement in studying apoptosis was serendipitous. However, it is clear now that this type of cell death may play a pivotal role in growth regulation in uterine epithelium as well as in other epithelia. Its occurrence in skin, intestine, etc., has already been described (Kerr and Harmon, this volume).

Our data clearly show the coordinated but inverse regulation of cell proliferation and death (Lynch et al. 1986; Nawaz et al. 1987; Rotello et al. 1991b). The simple equation

[GROWTH] = [CELL PROLIFERATION] – [CELL DEATH]

becomes essential for the understanding not only of normal tissue growth, but also of neoplasia. Figure 3 shows diagrammatically the complex regulatory mechanisms involved in growth regulation. All are essential parts of the schematic equation given above.

It has been shown that many tumors grow not only by increased cell proliferation, but also by decreased rate of cell loss (Baserga and Wiebel 1969). Genes from cancer cells appear to extend the survival of transfected normal cells, which was proposed to be a first step for neoplastic transformation (Vaux et al. 1988). This concept also becomes important for the design of cancer treatments, since most of the accepted cancer therapies rely on the idea that cancer cells proliferate at a faster rate with a concomitant increase in the rate of DNA synthesis (Lane 1977). Interestingly, Kerr et al. (1972) proposed early on that apoptosis could play an opposite role to mitosis in tissue size regulation.

Apoptosis appears to be regulated by either steroid or peptide hormones (Fig. 4). Glucocorticoid addition induces it in lymphoid cells (Wyllie 1980; Compton and Cidlowski 1986), and removal of interleukin-2 has a similar effect on IL-2-dependent cells (Duke and Cohen 1986). Withdrawal of progesterone elicits apoptosis in uterine epithelial cells (Rotello et al. 1989, 1991a), lack of erythropoietin evokes it in bone marrow cells (Koury and Bondurant 1990), and removal of androgens induces it in prostatic epithelium (Kyprianou and Isaacs 1988; Buttyan et al. 1989). On the other hand, addition

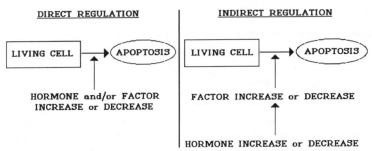

FIGURE 4 It is proposed that apoptosis is triggered by either direct or indirect means.

of rTGF-β1 or rMIS results in increased apoptosis in cultured endometrial cells (Rotello et al. 1990, 1991b). Therefore, it appears that the regulation of apoptosis can be due to either addition or withdrawal of hormones or factors. It is not yet clear if there is a causal relationship between hormones working in endocrine fashion versus paracrine factors, but data have been published suggesting the possibility that androgens may be regulating apoptosis through TGF-β1 (Kyprianou and Isaacs 1989) and that this peptide also induces involution in mammary gland (Silberstein et al. 1990), which is under stringent steroid hormone growth control.

Research using inhibitors of transcription and protein synthesis has suggested that both pathways are necessary for apoptosis (Wyllie et al. 1984; Martin et al. 1988), implying that an intracellular positive signal may be necessary (Fig. 5). However, recent data have shown that those inhibitors can induce apoptosis (Martin et al. 1990), suggesting that a short half-life

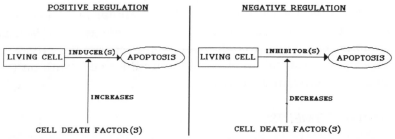

FIGURE 5 It is proposed that apoptosis is regulated by intracellular inducer(s) and/or inhibitor(s).

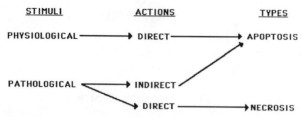

FIGURE 6   Classification of cell death. Pathological stimuli or injury (harmful to the host) will result in necrosis when acting in direct fashion. However, it may result in apoptosis either if it mimics physiological stimuli or if the latter is triggered by a possible repair process. The definition of apoptosis vs. necrosis involves morphological as well as biochemical analysis of DNA.

repressor of apoptosis may be present in cells and that exogenous induction could result in inactivation of such a molecule(s). These conceptual differences are important for the molecular biological study of apoptosis: Are there apoptosis or cell death genes? Significant work in this area has been done with the nematode *Caenorhabditis elegans* (Ellis and Horvitz 1986; Yuan and Horvitz 1990) but has not yet been extended successfully to mammalian systems. It is possible that apoptosis regulation is elicited differently in various biological systems.

CONCLUSIONS

It is postulated that apoptosis is a normal and important component of the growth process. It appears to be regulated in a coordinated but inverse fashion with cell proliferation. An increased internucleosomal DNA breakdown and endonuclease activity appear to be associated with the morphological changes. It is not yet known what the cellular and molecular mechanisms involved in apoptosis are. What is clear is that this form of cell death is different from necrosis. We proposed the simple classification given in Figure 6 for cell death on the basis of the current evidence.

ACKNOWLEDGMENTS

Our research has been supported throughout the years by the National Institutes of Health, the National Cancer Institute, and the American Cancer Society. Departmental colleagues have helped with suggestions and criticism. We thank Ms. Rita Lieberman for excellent experimental help, Ms. Laurie Bogue for typing this manuscript, Drs. Ken Shroyer and Peter Lapis for reviewing it, and Ms. Mariana Gerschenson-LiButti for helping with the figures.

REFERENCES

Arends, M.J., R.G. Morris, and A.H. Wyllie. 1990. Apoptosis: The role of the endonuclease. *Am. J. Pathol.* **136:** 593.

Baba, N. and E. von Haam. 1967. Experimental carcinoma of the endometrium. *Prog. Exp. Tumor Res.* **9:** 192.

Baserga, R. and F. Wiebel. 1969. The cell cycle of mammalian cells. *Int. Rev. Exp. Pathol.* **7:** 1.

Baulieu, E.E. 1989. Contragestion and other clinical applications of RU 486, an antiprogesterone at the receptor. *Science* **245:** 1351.

Beier, H.M. 1968. Uteroglobin: A hormone sensitive endometrial protein involved in blastocyst development. *Biochim. Biophys. Acta* **160:** 289.

Buttyan, R., C.A. Olsson, J. Pintar, C. Chang, M. Bandyk, P. Ng, and I.S. Sawczuk. 1989. Induction of the TRPM-2 gene in cells undergoing programmed death. *Mol. Cell. Biol.* **9:** 3473.

Compton, M.M. and J.A. Cidlowski. 1986. Rapid *in vivo* effects of glucocorticoids on the integrity of rat lymphocyte genomic deoxyribonucleic acid. *Endocrinology* **118:** 38.

Conti, C.J., I. Gimenez-Conti, and L.E. Gerschenson. 1981. Differential effects of 17β-estradiol and progesterone on the proliferation of glandular and luminal cells of rabbit uterine epithelium. *Biol. Reprod.* **24:** 643.

Conti, C.J., I. Gimenez-Conti, E.A. Conner, J.M. Lehman, and L.E. Gerschenson. 1984. Estrogen and progesterone regulation of proliferation, migration and loss of different target cells of rabbit uterine epithelium. *Endocrinology* **114:** 345.

Duke, R.C. and J.J. Cohen. 1986. IL-2 addiction: Withdrawal of growth factor activates a suicide program in dependent T cells. *Lymphokine Res.* **5:** 289.

Ellis, H.M. and H.R. Horvitz. 1986. Genetic control of programmed cell death in the nematode *C. elegans*. *Cell* **44:** 817.

Gerschenson, L.E. and J.A. Berliner. 1976. Further studies on the regulation of cultured rabbit endometrial cells by diethylstilbestrol and progesterone. *J. Steroid Biochem.* **7:** 159.

Gerschenson, L.E., J.A. Berliner, and J.J. Yang. 1974. Diethylstilbestrol and progesterone regulation of cultured rabbit endometrial cell growth. *Cancer Res.* **34:** 2873.

Gerschenson, L.E., E.A. Conner, and J.T. Murai. 1977. Regulation of the cell cycle by diethylstilbestrol and progesterone in cultured endometrial cells. *Endocrinology* **100**: 1468.

Gerschenson, L.E., J.R. Depaoli, and J.T. Murai. 1981. Inhibition of estrogen-induced proliferation of cultured rabbit uterine epithelial cells by a cell density-dependent factor produced by the same cells. *J. Steroid Biochem.* **14**: 959.

Gerschenson, L.E., J. Gorski, and D.M. Prescott. 1984. Induction of DNA synthesis in cultured rabbit uterine cells by estradiol and inhibition of the estrogen response. *J. Steroid Biochem.* **21**: 135.

Gerschenson, L.E., E.A. Conner, J. Yang, and M. Anderson. 1979. Hormonal regulation of proliferation in two populations of rabbit endometrial cells in culture. *Life Sci.* **24**: 1337.

Greenblatt, R.B., R.D. Gambrell, and L.D. Stoddard, Jr. 1982. The protective role of progesterone in the prevention of endometrial cancer. *Pathol. Res. Pract.* **174**: 297.

Greene, H.S.N. 1959. Adenocarcinoma of the uterine fundus in the rabbit. *Ann. N.Y. Acad. Sci.* **75**: 535.

Griffiths, G.D., M.D. Leek, and D.J. Gee. 1987. The toxic plant proteins ricin and abrin induce apoptotic changes in mammalian lymphoid tissues and intestine. *J. Pathol.* **151**: 221.

Guiochon-Mantel, A., H. Loosfelt, T. Ragot, A. Bailly, M. Atger, M. Misrahi, M. Perricaudet, and E. Milgrom. 1988. Receptors bound to antiprogestin form abortive complexes with hormone responsive elements. *Nature* **336**: 695.

Hewish, D.R. and L.A. Burgoyne. 1973. Chromatin substructure. The digestion of chromatin DNA at regularly spaced sites by a nuclear deoxyribonuclease. *Biochem. Biophys. Res. Commun.* **52**: 504.

Hopwood, D. and D.A. Levison. 1976. Atrophy and apoptosis in the cyclical human endometrium. *J. Pathol.* **119**: 159.

Keenan, K.P., D.D. Sharpnack, H. Collins, S.D. Formal, and A.D. O'Brien. 1986. Morphologic evaluation of the effects of Shiga toxin and *E. coli* Shiga-like toxin on the rabbit intestine. *Am. J. Pathol.* **125**: 69.

Kerr, J.F.R., A.H. Wyllie, and A.R. Currie. 1972. Apoptosis: A basic biological phenomenon with wide-ranging implications in tissue kinetics. *Br. J. Cancer* **26**: 239.

Koury, M.J. and M.C. Bondurant. 1990. Erythropoietin retards DNA breakdown and prevents programmed death in erythroid progenitor cells. *Science* **248**: 378.

Kyprianou, N. and J.T. Isaacs. 1988. Activation of programmed cell death in the rat ventral prostate after castration. *Endocrinology* **122**: 552.

————. 1989. Expression of transforming growth factor-β in the rat ventral prostate during castration-induced programmed cell death. *Mol. Endocrinol.* **3**: 1515.

Lane, M. 1977. Cancer chemotherapy. In *Cancer*, 5th. edition (ed.

J.A. del Regato and H.J. Spjut), p. 105. Mosby, St. Louis, Missouri.

Low, R.L., J.M. Buzan, and C.L. Couper. 1988. The preference of the mitochondrial endonuclease for a conserved sequence block in mitochondrial DNA is highly conserved during mammalian evolution. *Nucleic Acids Res.* **16:** 6427.

Lynch, M., S. Nawaz, and L.E. Gerschenson. 1986. Evidence for soluble factors regulating cell death and cell proliferation in primary cultures of rabbit endometrial cells grown on collagen. *Proc. Natl. Acad. Sci.* **83:** 4784.

Martin, D.P., R.E. Schmidt, P.S. DiStefano, O.H. Lowry, J.G. Carter, and E.M. Johnson. 1988. Inhibitors of protein synthesis and RNA synthesis prevent neuronal death caused by nerve growth factor deprivation. *J. Cell Biol.* **106:** 829.

Martin, S.J., S.V. Lennon, A.M. Bonham, and T.G. Cotter. 1990. Induction of apoptosis (programmed cell death) in human leukemic HL-60 cells by inhibition of RNA or protein synthesis. *J. Immunol.* **145:** 1859.

Murai, J.T., C.J. Conti, I. Gimenez-Conti, D. Orlicky, and L.E. Gerschenson. 1981. Temporal relationship between rabbit uterine epithelium proliferation and uteroglobulin production. *Biol. Reprod.* **24:** 649.

Nawaz, S., P. Galand, and L.E. Gerschenson. 1987. Hormonal regulation of cell death in rabbit uterine epithelium. *Am. J. Pathol.* **127:** 51.

Rajkumar, K., R. Bigsby, R. Lieberman, and L.E. Gerschenson. 1983a. Uteroglobin production by cultured rabbit uterine epithelial cells. *Endocrinology* **112:** 1490.

———. 1983b. Effect of progesterone and 17β-estradiol on the production of uteroglobin by cultured rabbit uterine epithelial cells. *Endocrinology* **112:** 1499.

Roberts, A.B., K.C. Flanders, P. Kondaiah, N.L. Thompson, E. van Obberghen-Shilling, L. Wakefield, P. Rossi, B. de Crombrugghe, V. Heine, and M.B. Sporn. 1988. Transforming growth factor β: Biochemistry and role in embryogenesis, tissue repair and remodeling, and carcinogenesis. *Recent Prog. Horm. Res.* **44:** 157.

Rotello, R.J., M.B. Hocker, and L.E. Gerschenson. 1989. Biochemical evidence for programmed cell death in rabbit uterine epithelium. *Am. J. Pathol.* **134:** 491.

Rotello, R.J., R.C. Lieberman, and L.E. Gerschenson. 1991a. Characterization of uterine epithelium apoptotic cell death kinetics and regulation by progesterone and RU 486. *Am. J. Pathol.* (in press).

Rotello, R.J., R.C. Lieberman, A.F. Purchio, and L.E. Gerschenson. 1991b. Coordinated regulation of apoptosis and cell proliferation by TGF-β1 in cultured cells. *Proc. Natl. Acad. Sci.* **88:** 3412.

Rotello, R.J., R.L. Cate, R.C. Lieberman, and L.E. Gerschenson. 1990. Müllerian inhibiting substance increases apoptosis in uterine epithelium primary cultures. *J. Cell Biol.* **111:** 477a.

Shroyer, K.R., R.C. Lieberman, and L.E. Gerschenson. 1987a. Estradiol-17β and progesterone regulate secretion of uteroglobin through different pathways. *Histochemistry* **87:** 173.

Shroyer, K.R., C.L. Williams, G.J. Miller, and L.E. Gerschenson. 1987b. Uteroglobin production in the pseudopregnant rabbit uterus. Immunohistochemical studies. *Histochemistry* **87:** 471.

Silberstein, G.B., P. Strickland, S. Coleman, and C.W. Daniel. 1990. Epithelium-dependent extracellular matrix synthesis in transforming growth factor-β1-growth-inhibited mouse mammary gland. *J. Cell Biol.* **110:** 2209.

Siiteri, P.K. 1978. Steroid hormones and endometrial cancer. *Cancer Res.* **38:** 4360.

Vanderbilt, J.N., K.S. Bloom, and J.N. Anderson. 1982. Endogenous nuclease: Properties and effects on transcribed genes in chromatin. *J. Biol. Chem.* **257:** 13009.

Vaux, D.L., S. Cory, and J.M. Adams. 1988. Bcl-2 gene promotes haemopoietic cell survival and cooperates with c-*myc* to immortalize pre-B cells. *Nature* **355:** 440.

Wyllie, A.H. 1980. Glucocorticoid-induced thymocyte apoptosis is associated with endogenous endonuclease activation. *Nature* **284:** 555.

Wyllie, A.H., R.G. Morris, A.L. Smith, and D. Dunlop. 1984. Chromatin cleavage in apoptosis: Association with condensed chromatin morphology and dependence on macromolecular synthesis. *J. Pathol.* **142:** 67.

Yuan, J. and H.R. Horvitz. 1990. The *Caenorhabditis elegans* genes *ced-3* and *ced-4* act cell autonomously to cause programmed cell death. *Dev. Biol.* **138:** 33.

Apoptotic Process in the Radiation-induced Death of Lymphocytes

S.R. Umansky

Institute of Biological Physics, USSR Academy of Sciences, Pushchino
Moscow region, 142292, Union of Soviet Socialist Republics

All primary syndromes of radiation disease are a consequence of radiation-induced cell death in sensitive tissues, which is the reason this phenomenon has been studied so intensively by radiobiologists during the past several decades. Two types of radiation-induced cell death are usually distinguished (Okada 1970), reproductive and interphase. Reproductive cell death occurs after one or even several divisions and is characteristic of actively proliferating populations. On the contrary, interphase cell death occurs before the next division and is specific for cells that are in a nonproliferative state at the time of irradiation. Although proliferating cells can die before the next mitosis, this usually occurs following extremely high doses of radiation.

In general, cells vary in their radiosensitivity, and two types of cells can be distinguished with respect to interphase death. Sensitive cells, such as lymphocytes, die following as little as ten to as much as several hundred rads, whereas resistant cells, such as hepatocytes, can survive several thousand rads. Reproductive cell death is considered to be caused by radiation-induced chromosomal aberrations with partial inactivation or loss of genetic material. However, decades of investigation into the mechanism of interphase cell death has led to the conclusion that this mode of death is fundamentally different from that of reproductive death. During the 1950s and the 1960s, a great deal of data was gathered on the morphological and biochemical characteristics of interphase death. However, the significance of these data was not clear until late in the next decade.

DNA Degradation in Dying Lymphocytes

In 1957, Cole and Ellis showed the dose-dependent accumulation of soluble polydeoxyribonucleotides in spleen and bone marrow of irradiated animals and suggested its possible importance in the interphase cell death. Subsequently, this phenomenon was further investigated in several other radiobiological laboratories (Skalka and Matyasova 1963; Vodolazskaya and Yermolaeva 1971; Yermolaeva and Vodolazskaya 1977; Ivannik et al. 1978), where it was shown that polydeoxyribonucleotides are products of chromatin degradation and appear as complexes of DNA with histones and nonhistone proteins. Such chromatin degradation products were found to be formed in lymphocytes not only after irradiation, but also following several other kinds of cytotoxic stress such as glucocorticoid treatment. The kinetics of the appearance of chromatin fragmentation was found to be similar to that of cell death as measured cytomorphologically by the appearance of pycnosis.

Skalka et al. (1976) showed that the DNA of polydeoxyribonucleotides was similar in length to DNA of nucleosomes and their oligomers. Subsequently, detailed investigations have revealed that the products of chromatin degradation are actually nucleosomes and their oligomers with normal content of histones and nonhistone proteins. This observation revealed that fragmentation of the genetic material is not simply a consequence of autolytic digestion of cell components by lysosomal enzymes (Umansky et al. 1980, 1981a, 1988; Zhivotovsky et al. 1981). Furthermore, these results were in accord with morphological studies by Wyllie et al. (1973, 1980) on radiation-induced death of lymphocytes; this type of cell death was called *apoptosis* by these investigators.

Flow cytofluorometry has made it possible to analyze DNA degradation in individual cells (Afanasyev et al. 1986; Pechatnikov et al. 1986). Using this approach, cells can be fixed and stained with a DNA-specific dye under conditions that permit the extraction of chromatin degradation products. Cells that contain fragmented DNA can then be detected as those containing less than 2C DNA (Fig. 1). In our institute, using freshly prepared thymocytes, chromatin degradation was found to begin after a 1–2-hour lag period following irradiation and

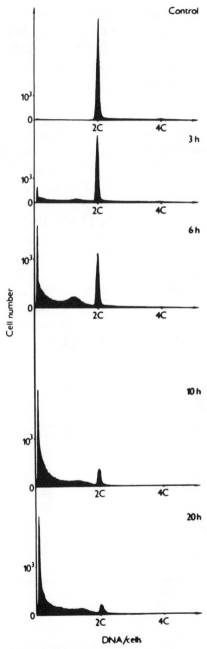

FIGURE 1 DNA histograms of thymocytes from γ-irradiated (10 Gy) rats, stained with Hoechst 33258. Time after irradiation is indicated on the figure.

reached maximal values after 8–10 hours. Flow cytofluorometry further revealed that DNA is cleaved in each dying
thymocyte very rapidly, and the progressive total accumulation
of chromatin fragments with time reflects the kinetics of initiation of new cells in that death process, but not deepening of
DNA degradation in individual cells.

Analysis of the kinetics of DNA degradation demonstrated
the complex nature of the physiological process. It was found
that the ratio of mononucleosomes to oligomers of different
lengths did not depend on the dose or the length of time after
irradiation (Umansky et al. 1981a,b). The same characteristics
were observed for degradation of individual genes (Umansky et
al. 1988; Beletsky et al. 1989a). The transcribed and potentially inducible genes are cleaved to nucleosomes and their oligomers in thymocytes of irradiated animals. On the other hand,
albumin gene, which is not transcribed in this tissue, is also
degraded, but no low-molecular-weight fragments were found
(Fig. 2). In both instances, the degree of gene degradation was
found to be invariable in time. Nucleases appeared to be active
for a short time, and it appeared that the degree of DNA frag-

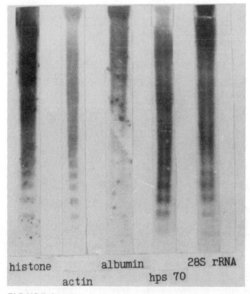

histone   albumin   28S rRNA
      actin   hps 70

FIGURE 2 Blot-hybridization of DNA isolated from rat thymocytes 4 hr
after γ-irradiation with the sequences of various genes.

mentation could be dependent on either chromatin structure or the subnuclear distribution of endonucleolytic enzyme activity.

DNA degradation is characteristically observed to precede an increase in plasma membrane permeability (Fig. 3A), which has been the most commonly accepted biochemical criterion of apoptosis. However, the observation of internucleosomal chromatin cleavage, in combination with specific cellular ultrastructural changes (see Kerr and Harmon, this volume), is the most widely accepted practical basis on which to define cell death. Conversely, necrotic cell death has been character-

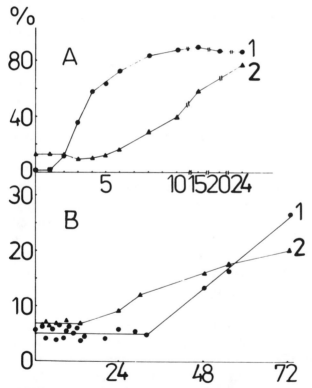

FIGURE 3 The γ-irradiation-induced apoptosis of thymocytes (A) and necrosis of Burkitt's lymphoma cells (B). Appearance of cells with degraded DNA (1) and those permeable for ethidium bromide (2) was measured by flow cytometry. The hours after exposure are given on the abscissa.

ized by an early increase in plasma membrane permeability with a subsequent DNA degradation that yields irregular fragment sizes (Fig. 3B), probably a result of the activation of hydrolases in the already dead cells (Afanasyev et al. 1986).

Ca⁺⁺,Mg⁺⁺-dependent Nuclear Nuclease Is the Enzyme Degrading DNA in Dying Cells

The repair of direct radiation-induced DNA breaks is normally complete within 30 minutes in thymocytes. However, DNA cleavage in thymocytes of γ-irradiated or hydrocortisone-treated animals begins with accumulation of 3'-OH single-strand breaks in the internucleosomal linkers 40–60 minutes after their treatment. The single-strand breaks are then converted into double-strand breaks (Beletsky and Umansky 1987; Umansky et al. 1988), which then give rise to characteristic DNA ladder formation following agarose gel electrophoresis. Analysis of nucleases that are active in isolated thymocyte nuclei have also shown that this type of DNA degradation is characteristic of Ca⁺⁺,Mg⁺⁺-dependent nuclease (Fig. 4).

Nuclear enzyme activity was shown to increase threefold in nuclei isolated 2 hours after irradiation, and sevenfold in nuclei isolated at 3 hours (Fig. 5) (Nikonova et al. 1982). Ca⁺⁺,Mg⁺⁺-dependent nuclease activity is inhibited by Zn⁺⁺ addition, and addition of Zn⁺⁺ to irradiated or glucocorticoid-treated thymocytes also blocked DNA fragmentation (Cohen and Duke 1984; Beletsky et al. 1989b). Cycloheximide introduction strongly inhibits the DNA fragmentation by Ca⁺⁺,Mg⁺⁺-dependent nucleases in isolated nuclei. Enzyme activity was found to have diminished by 80% at 2 hours after cycloheximide treatment, with negligible activity observable at 3 hours (Nikonova et al. 1982). These data strongly suggest that a Ca⁺⁺,Mg⁺⁺-dependent nuclease or a nuclease-activating protein turns over very rapidly in these systems and that the cell death is dependent on the continuing synthesis of either enzyme or enzyme-activating protein.

Data have clearly indicated the involvement of a Ca⁺⁺,Mg⁺⁺-nuclease in the degradation of DNA in dying cells. Although the mechanisms of this process are not yet clear, some general mechanisms have been considered by our laboratory.

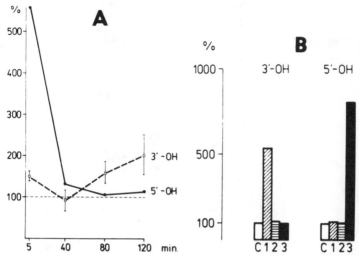

FIGURE 4 Kinetics of the formation of single-strand breaks, bearing 3'-OH and 5'-OH ends, in thymocytes of γ-irradiated (10 Gy) rats (A), and the induction of 3'-OH and 5'-OH end breaks by autolytic DNA cleavage in thymocyte nuclei (B). The number of 3'-OH and 5'-OH ends was measured by nick translation and direct polynucleotide kinase reaction, respectively. The ordinates represent label incorporation as a percentage of the control. (A) The abscissa represents the time after irradiation. (B) (C) Control; (1) Ca^{++},Mg^{++}-nuclease; (2) Mg^{++}-nuclease; (3) acidic nuclease.

Change in the structure of chromatin. Altered chromatin structure could lead to enhanced sensitivity to preexisting nucleolytic enzyme. Such a structural change could be caused by limited proteolysis or other related events. Indeed, an increased accessibility of antigenic determinants in thymocyte chromatin 3 hours after irradiation has been found, which indicates a decrease in chromatin condensation (Zotova et al. 1985). However, the sensitivity of similar thymocyte chromatin preparations to different enzymes, such as DNase I, micrococcal and *Serratia marcescens* nucleases, or endogenous acidic nuclease, does not differ from that of control preparations (Nikonova et al. 1982). However, one cannot exclude the possibility that the activity of the Ca^{++},Mg^{++}-dependent nuclease is more sensitive to chromatin structure than is that of these nucleases.

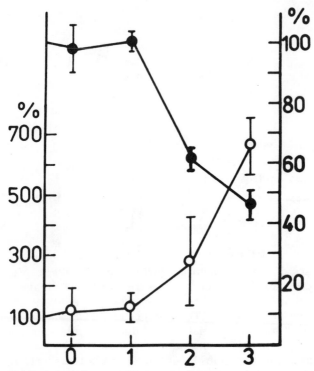

FIGURE 5 Influence of γ-irradiation (10 Gy) of rats on the activity of poly(ADP-ribose) polymerase (filled circle) and Ca^{++},Mg^{++}-nuclease (open circle) in isolated thymocyte nuclei. The abscissa represents the time after irradiation (hr). The right and left ordinates represent the activity of poly(ADP-ribose) polymerase and nuclease, respectively.

De novo synthesis of specific nuclease. The Ca^{++},Mg^{++}-dependent nuclease activity can be observed in nuclei of normal thymocytes and some other nondividing cells; however, similar nuclease activity is usually absent in proliferating cells (Cohen and Duke 1984). In proliferating cells, this enzyme activity has been shown to appear following treatments that cause DNA degradation and subsequent cell death. Nonetheless, this observation does not necessarily mean that actively proliferating cells are devoid of the nuclease. Rather, it is more likely that a preexisting enzyme may be reversibly inhibited in view of the observation that, in some experimental systems, induction of Ca^{++},Mg^{++}-dependent nuclease activity does not depend on new protein synthesis. Hence, in at least some cell types, it is not likely that the availability of the enzyme is a limiting step in DNA degradation in apoptosis. Final clarification of this mechanism may depend on availability of specific antibodies against Ca^{++},Mg^{++}-dependent nuclease, as well as a probe for its corresponding gene.

Activation of the preexisting enzyme. There is some evidence that the specific DNA fragmentation observed during apoptosis is the consequence of activation of a preexisting nuclear enzyme. It has been shown that Ca^{++},Mg^{++}-dependent nuclease is inhibited by poly(ADP-ribosylation). Furthermore, severe depression of the poly(ADP-ribose) polymerase activity in irradiated thymocytes has been observed preceding the onset of DNA degradation (Fig. 5) (Nelipovich et al. 1988). Therefore, it is conceivable that normally the newly synthesized molecules of the nuclease undergo poly(ADP-ribosylation) that results in its inactivation. Following γ-irradiation, the activity of poly(ADP-ribose) polymerase decreases, and de novo synthesized Ca^{++},Mg^{++}-dependent nuclease, which turns over rapidly, retains high activity and cleaves DNA.

Nuclease activation by modification, e.g., restricted proteolysis. Using SDS-PAGE with subsequent analysis of nuclease activity, we have observed that the molecular weight of a substantial portion of Ca^{++},Mg^{++}-dependent nuclease decreases 2 hours after irradiation. Of course, other possibilities cannot be excluded.

Dependence of DNA Degradation on Protein Synthesis

In many instances, DNA degradation and other manifestations of apoptotic cell death can be prevented by inhibitors of protein or RNA synthesis (Yermolaeva and Vodolazskaya 1977; Ivannik et al. 1978; Umansky et al. 1980). Usually this fact is interpreted as an indication of the necessity of new gene induction. However, there are other possible mechanisms.

In apoptotic cells, rapidly turning over proteins, e.g., Ca^{++},Mg^{++}-dependent nuclease, which take part in the DNA fragmentation, diminish upon the treatment with inhibitors of protein synthesis, and there is no enzyme activity that can cleave DNA by the necessary time. On the other hand, the appearance of new proteins preceding DNA cleavage has been shown in different systems (Wadewitz and Lockshin 1988; Zhivotovsky et al. 1988; Domashenko et al. 1990). Unfortunately, it is quite difficult to prove in each particular case that the new proteins actually participate in apoptosis. In our study, this possibility is indirectly supported by the fact that two cytotoxic agents, irradiation and glucocorticoids, acting through different pathways, induce synthesis of the same protein in thymocytes. However, in other experimental systems, e.g., the cell death induced by cytotoxic lymphocytes or natural killer cells, apoptosis does not depend on protein synthesis (Duke et al. 1983). Moreover, cell death caused by tumor necrosis factor (TNF) was facilitated by inhibitors of protein or RNA synthesis (Rubin et al. 1988). Some established cell lines that are normally insensitive to TNF lose their resistance in the presence of cycloheximide. It is likely that there are two groups of proteins in a cell. Activation of the first group is a necessary step of apoptosis, and the second group is its antagonist. The correlation of these systems and the effect of protein synthesis inhibitors are determined by the state of cell differentiation and by the stage of cell cycle. From this point of view, it is very interesting that diphtheria toxin, which induces cell death by apoptosis, possesses two activities: It is a translation inhibitor and nuclease (Chang et al. 1989). Radiation and glucocorticoids depress the synthesis of specific proteins in thymocytes (Domashenko et al. 1990), as do many other cytotoxic agents. Thus, it is possible that the prevalence of the first system over the second one is a necessary condition of DNA fragmentation and cell death. In some cases, the synthesis of specific protein should be induced, but for other systems, it is important to repress the activity or synthesis of antagonistic proteins. Perhaps in most instances, apoptotic DNA degradation requires both the activation of the one group of proteins and the repression of the others.

Relationship between Cell Death and Proliferation

In contrast to lymphocytes, irradiated proliferating cells commonly die by necrosis (Afanasyev et al. 1986). However, recently it was shown that some glucocorticoid-dependent thymoma cell lines die by apoptosis upon γ-irradiation or hormone treatment in which internucleosomal DNA fragmentation is observed to precede the increase of cell membrane permeability. In these cells, there was an initial delay in cell cycle progression in G_2/M, and death occurred only after the blockade had been overcome and cytokinesis was restored (Figs. 6 and 7). The cells that were irreversibly blocked in G_2 were found to die by necrosis and not by apoptosis (Afanasyev et al. 1991). It is interesting that hormone removal 2–10 hours after treatment prevented cell death, but blockade in G_2 was still observed. Reduction of serum concentration in both irradiated and dexamethasone-treated cultures caused a shortening of the lag that was noted to precede DNA fragmentation. Presumably, the cells underwent a reprogramming from proliferation to death during the G_2 delay. It is possible that such a reprogramming can also take place in cells delayed in G_1. This supposition is in accord with the data on cell death caused by withdrawal of the growth factors or hormones. In any case, the relationship between cell death and proliferation is an intriguing problem, since the disturbance of the correlation can play an important role in carcinogenesis.

CONCLUDING REMARKS

The main achievement in the problem discussed seems to be a new point of view on the mechanisms of cell death. Now it is clear that, in most cases, death is an active cell response to physiological or damaging signals, DNA degradation being the key step of the preceding processes (Hanson 1979; Wyllie et al. 1980; Umansky 1982). Now there are only indirect data indicating the requirement for the induction of new genes taking part in the realization of the cell death program. The direct answer can be obtained in experiments on initiation of cell death by RNA isolated, e.g., from irradiated thymocytes, although the choice of adequate recipient cells is not a simple problem.

The initial and intermediate steps of apoptosis are practically unknown, although some data on possible involvement of Ca^{++} have been published recently (Orrenius et al. 1989). It is

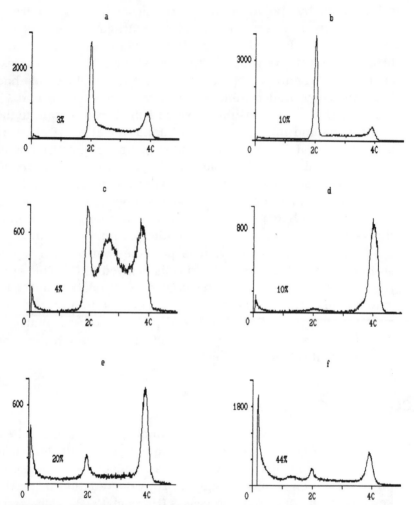

FIGURE 6 DNA-histograms of control (*a,b*) and γ-irradiated (*c–f*) thymoma BW-5147 cells, stained with Hoechst 33258. Time of incubation: (*a*) 0 hr, (*b*) 48 hr, (*c*) 6 hr, (*d*) 12 hr, (*e*) 24 hr, (*f*) 48 hr. The percentage of dying cells is indicated on each histogram. The abscissa is the amount of DNA per cell. The ordinate is the number of cells.

not completely clear why cells die soon after DNA fragmentation. It could be a result of deficiency in genetic information, but the same cells do not die for a longer time in the presence of actinomycin D. Although the toxic effect of the chromatin degradation products cannot be excluded, probably DNA deg-

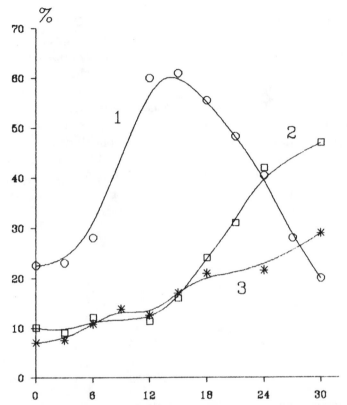

FIGURE 7 Effect of γ-irradiation (10 Gy) on thymoma BW-5147 cells. (1) Accumulation of cells in G_2/M. (2,3) Appearance of cells with degraded DNA and those permeable for trypan blue, respectively. Hours after exposure are given on the abscissa, and the percentage of cells is given on the ordinate.

radation is not a single irreversible event of cell death program; e.g., DNA cleavage and increased membrane permeability, induced by cytotoxic lymphocytes, seem to be independent processes. Perhaps understanding of mechanisms of membrane blebbing and apoptotic body formation will clarify the question.

One of the most serious radiobiological problems is the modification of the radiation effects, in particular, cell death. The new point of view on the mechanisms of cell death indicates the possibility of its modification using the principally new approaches: inhibition of activity of specific nucleases or other proteins involved in the process and so on. On the other hand, the question arises about the possible consequences of such a modification. It was supposed that one function of the cell death program is to eliminate the damaged cells; in particular, cells with oncogenic features (Umansky 1982). If this is so, the prevention of cell death can cause the increase of a carcinogenic risk. This example indicates the practical importance of the investigation of the mechanisms of cell death.

In a short review, it is impossible to touch all aspects of the problem of cell death. Obviously, the problem is far from being completely solved, but today's great interest promises progress in the near future.

REFERENCES

Afanasyev, V.N., B.A. Korol', Y.A. Mantsygin, P.A. Nelipovich, V.A. Pechatnikov, and S.R. Umansky. 1986. Flow cytometry and biochemical analysis of DNA degradation characteristic of two types of cell death. *FEBS Lett.* **194:** 347.

Afanasyev, V.N., B.A. Korol', I.I. Kruman, N.P. Matylevich, V.A. Pechatnikov, and S.R. Umansky. 1991. Relationship between cell proliferation and cell death. In *Modulating factors in multistage chemical carcinogenesis* (ed. A. Columbano and G.M. Ledda-Columbano). Plenum Press, New York. (In press.)

Beletsky, I.P. and S.R. Umansky. 1987. The nature of breaks induced in DNA at early stages of interphase cell death. *Radiobiologiya* **27:** 227.

Beletsky, I.P., A.V. Lichtenstein, and S.R. Umansky. 1989a. Individual gene degradation in irradiated rat thymocytes. *Radiobiologiya* **29:** 435.

Beletsky, I.P., J. Maytasova, L.V. Nikonova, M. Skalka, and S.R. Umansky. 1989b. On the role of Ca, Mg-dependent nuclease in the postirradiation degradation of chromatin in lymphoid tissues. *Gen. Physiol. Biophys.* **8:** 381.

Chang, M.P., R.L. Baldwin, D. Bruce, and B.J. Wisnieski. 1989. Second cytotoxic pathway of diphtheria toxin suggested by nuclease activity. *Science* **246:** 1165.

Cohen, J.J. and R.C. Duke. 1984. Glucocorticoid activation of calcium-dependent endonuclease in thymocyte nuclei leads to cell death. *J. Immunol.* **132:** 38.

Cole, L.J. and M.E. Ellis. 1957. Radiation-induced changes in tissue nucleic acids: Release of soluble deoxypolynucleotides in the spleen. *Radiat. Res.* **7:** 508.

Domashenko, A.D., L.F. Nazarova, and S.R. Umansky. 1990. Comparison of the spectra of proteins synthesized in mouse thymocytes after irradiation or hydrocortisone treatment. *Int. J. Radiat. Biol.* **57:** 315.

Duke, R.C., R. Chervenak, and J.J. Cohen. 1983. Endogenous endonuclease-induced DNA fragmentation: An early event in cell-mediated cytolysis. *Proc. Natl. Acad. Sci.* **80:** 6361.

Hanson, P. 1979. Molecular mechanisms of lymphoid cell interphase death. *Radiobiologiya* **19:** 814.

Ivannik, B.P., R.V. Golubeva, V.I. Murzaev, and N.I. Ryabchenko. 1978. Role of protein synthesis in postirradiation degradation of DNP and DNA in rat thymocytes. *Radiobiologiya* **18:** 803.

Nelipovih, P.A., L.V. Nikonova, and S.R. Umansky. 1988. Inhibition of poly-(ADP-ribose)polymerase as a possible reason for activation of Ca^{++}/Mg^{++}-dependent endonuclease in thymocytes of irradiated rats. *Int. J. Radiat. Biol.* **53:** 749.

Nikonova, L.F., P.A. Nelipovich, and S.R. Umansky. 1982. The involvement of nuclease nucleases in rat thymocyte DNA degradation after γ-irradiation. *Biochim. Biophys. Acta* **699:** 281.

Okada, S. 1970. *Radiation biochemistry: Cells,* vol. 1, p. 247. Academic Press, New York.

Orrenius, S., D.J. McConkey, G. Bellomo, and P. Nictera. 1989. Role of Ca^{2+} in toxic cell killing. *Trends Pharmacol. Sci.* **10:** 281.

Pechatnikov, V.A., V.N. Afanasyev, B.A. Korol', V.N. Korneev, Y.A. Rochev, and S.R. Umansky. 1986. Flow cytometry analysis of DNA degradation in thymocytes of γ-irradiation or hydrocortisone treated rats. *Gen. Physiol. Biophys.* **5:** 273.

Rubin, B.Y., L.J. Smith, G.R. Hellermann, R.M. Lunn, N.K. Richardson, and S.L. Anderson. 1988. Correlation between the anticellular and DNA fragmenting activities of tumor necrosis factor. *Cancer Res.* **48:** 6006.

Skalka, M. and J. Matyášová. 1963. The effect of low radiation doses on the release of deoxyribopolynucleotides in hematopoietic and lymphatic tissues. *Int. J. Radiat. Biol.* **7:** 41.

Skalka, M., J. Matyášová, and M. Čejková. 1976. DNA in chromatin of irradiated lymphoid tissues degrades *in vivo* into regular fragments. *FEBS Lett.* **72:** 271.

Umansky, S.R. 1982. The genetic program of cell death. Hypothesis and some applications: transformation, carcinogenesis, ageing. *J. Theor. Biol.* **97:** 591.

Umansky, S.R., B.A. Korol', and P.A. Nelipovich. 1981a. In vivo DNA degradation in thymocytes of γ-irradiation or hydrocortisone treated rats. *Biochim. Biophys. Acta* **655:** 9.

Umansky, S.R., B.A. Korol', and P.A. Nelipovich. 1981b. In vivo DNA degradation in dying thymocytes—A model for studying of chromatin structure. *Mol. Biol. Mosc.* **15:** 1028.

Umansky, S.R., P.A. Nelipovich, B.A. Korol', and L.V. Nikonva. 1980.

Molecular mechanisms of thymocyte interphase death in gamma-irradiated rats. *Radiat. Environ. Biophys.* **17**: 325.

Umansky, S.R., I.P. Beletsky, B.A. Korol', A.V. Lichtenstein, and P.A. Nelipovich. 1988. Molecular mechanisms of DNA degradation in dying rodent thymocytes. *Mol. Cell. Biol.* **7**: 221.

Vodolazskaya, N.A. and N.V. Yermolaeva. 1971. Investigation of DNP degradation produces induced by γ-irradiation, hydrocortisone and degranol in rat thymus. *Radiobiologiya* **11**: 335.

Wadewitz, A.B. and R.A. Lockshin. 1988. Programmed cell death: Dying cells synthesize a co-ordinated, unique set of proteins in two different episodes of cell death. *FEBS Lett.* **241**: 19.

Wyllie, A.H., J.F.R. Kerr, and A.R. Currie. 1973. Cell death in the hormonal neonatal rat adrenal cortex. *J. Pathol.* **111**: 255.

―――. 1980. Cell death: The significance of apoptosis. *Int. Rev. Cytol.* **68**: 251.

Yermolaeva, N.V. and N.A. Vodolazskaya. 1977. A delay in the radiation-induced damage to lymphoid tissue caused by the action of protein synthesis inhibitors. *Radiobiologiya* **18**: 480.

Zhivotovsky, B.D., N.B. Zvonareva, and K.P. Hanson. 1981. Characteristics of rat thymus chromatin degradation products after whole-body X-irradiation. *Int. J. Radiat. Biol.* **39**: 437.

Zhivotovsky, B.D., L. Perlaky, A. Fonagy, and K.P. Hanson. 1988. Nuclear protein synthesis in thymocytes of X-irradiated rats. *Int. J. Radiat. Biol.* **54**: 999.

Zotova, R.N., L.F. Nazarova, V.M. Ermekova, and S.R. Umansky. 1985. Postirradiation changes in the chromatin structure of irradiated rat thymocytes. *Radiobiologiya* **25**: 517.

Apoptosis in Cell-mediated Immunity

R.C. Duke

Department of Microbiology and Immunology
Immunothanatology Unit
University of Colorado School of Medicine
Denver, Colorado 80262

Programmed Cell Death in the Immune System

The cells of the immune system provide numerous examples of programmed cell death (Cohen et al. 1985). In fact, it appears that the majority of leukocytes are extremely short-lived. The circumstances in which apoptosis may be induced in these cells are as diverse as the cells themselves. For example, immature T lymphocytes can be induced to undergo apoptosis upon exposure to glucocorticoids (Wyllie 1980; Vanderbilt et al. 1982; Cohen and Duke 1984), low-dose γ irradiation (Skalka et al. 1976; Zvonareva et al. 1983; Sellins and Cohen 1987), certain toxins including dioxin (Bell and Jones 1982; McConkey et al. 1988), heat shock (Sellins and Cohen 1991), and antibodies that cross-link their antigen receptors (McConkey et al. 1989; Shi et al. 1989; Smith et al. 1989). The mechanisms controlling programmed cell death in leukocytes are equally complex; apoptosis in cells of the immune system can be either dependent on, independent of, or induced by cessation of protein synthesis (Cohen 1991; Cohen et al. 1991).

It is beyond the scope of this paper to describe apoptosis in each of the various cell types; therefore, we concentrate on two examples that occur during a typical cell-mediated immune response. These examples are death of dependent T cells upon removal of growth factor (Bishop et al. 1985; Duke and Cohen 1986) and death of target cells induced by killer T cells (Battersby et al. 1974; Don et al. 1977; Duke et al. 1983, 1986; Russell 1983; Stacey et al. 1985).

Apoptosis: The Molecular Basis of Cell Death
Copyright 1991 Cold Spring Harbor Laboratory Press 0-87969-366-5/91 $3.00 + 00

Cell-mediated Immunity

For the purposes of this paper, we limit ourselves to a hypothetical T-cell-mediated immune response to a virus. However, much of what is described also pertains to the cells involved in antibody-mediated responses.

Our story begins in a lymph node where several virus particles have been ingested by a macrophage. Viral proteins are degraded within the macrophage's lysosomes, and polypeptides are generated. The virus-derived polypeptides (antigens) are expressed on the surface of the macrophage in association with major histocompatibility complex (MHC) molecules (Zinkernagel and Doherty 1974; Buus et al. 1985; Watts et al. 1985; Bjorkman et al. 1987). A virus-specific helper T lymphocyte (T_H) binds to the complex of viral antigen and MHC on the macrophage and becomes "activated." Once activated, the T_H cell begins to express and secrete lymphokines that have several effects; some lymphokines cause additional leukocytes to migrate to and remain in the lymph node, whereas others act as T-cell growth factors. These latter lymphokines include interleukin-2 (IL-2).

A virus-specific cytotoxic T lymphocyte (CTL) binds to the macrophage, which results in the expression of high-affinity receptors for IL-2 (Sharon et al. 1986; Weissman et al. 1986; Robb et al. 1987; Teshigawara et al. 1987). Low-affinity receptors for IL-2 are constitutively expressed by T cells, although their function is unclear. IL-2 secreted by the activated T_H cell binds to the high-affinity receptor on the CTL, which eventually proliferates and becomes able to mediate killing of virally infected cells.

During this "cell-mediated" immune response, the absolute number of virus-specific CTLs and T_H cells increases dramatically. The "swollen glands" associated with a sore throat provide a striking example of this proliferation. These glands are actually lymph nodes filled to their maximum volume with lymphocytes responding to the offending pathogen. What is the fate of these responding cells upon clearance of antigen? Results indicate that the majority of these growth-factor-dependent cells die via programmed cell death.

Apoptosis in IL-2-dependent T Cells upon Antigen Clearance

Gillis and Smith were among the first investigators to describe how antigen-specific CTLs could be maintained in vitro with IL-2 derived from T_H cells. In their initial report, they indicated that these CTLs were not merely responsive to the growth factor but were totally addicted, in that they died if IL-2 was withheld (Gillis et al. 1978). Why do IL-2-addicted T cells die?

Growth factors induce proliferation following activation of certain genes and the subsequent macromolecular synthesis necessary for cell growth and division (Cantrell and Smith 1984). Removal of growth factors would be expected to inhibit proliferation; however, the majority of antigen-specific cells actually die rather than become quiescent (Gillis et al. 1978). Cells deprived of growth factors should die eventually from a lack of appropriate stimulation. The biochemical characteristics of growth-factor-withdrawal-induced death suggested that apoptosis, rather than acquiescent expiration, was involved and that the dying cells were actively participating in their own suicide (Table 1).

TABLE 1 *CHARACTERISTICS OF DEATH OF IL-2-DEPENDENT MOUSE T CELLS UPON WITHDRAWAL OF GROWTH FACTOR*

1. Nuclear damage, as determined either by chromatin condensation by light microscopy; lack of sedimentation of chromatin following detergent lysis of cells in hypotonic buffer containing EDTA; or the appearance of oligonucleosome-sized fragments upon agarose gel electrophoresis of cellular DNA, occurs within *6–24 hr* of removal of IL-2.
2. Cell death, as measured by plasma membrane breakdown, follows the induction of nuclear damage by *4–8 hr.*
3. Nuclear damage and cell death *require de novo* protein synthesis.
4. Endogenous endonuclease activity *cannot* be detected in the nuclei of any of the IL-2-dependent T cells before or following growth factor withdrawal.

Summary of results presented in Duke and Cohen (1986) and Bishop et al. (1985).

The first cells tested were the CTLL-2 and HT-2 cell lines, which were derived from antigen-specific T-cell clones that had lost the ability to respond to antigen but grew continuously in the presence of exogenously added growth factor. This growth pattern is unlike that of more typical T cells, which require continued antigenic stimulation in order to maintain sufficient levels of expression of high-affinity IL-2 receptors. CTLL-2 and HT-2 cells are exquisitely sensitive to growth factor removal and begin to undergo apoptosis as early as 6 hours after withdrawal. Antigen-specific T cells, in contrast, seem to be slightly more recalcitrant to IL-2-deprivation and do not begin to fragment their DNA, and die, until after 24 hours or longer.

In both types of IL-2-dependent T cells, growth factor seems to both promote proliferation and prevent induction of apoptosis. This is best illustrated with the antigen-specific T cells. As mentioned above, antigenic stimulation is required for expression of high-affinity IL-2 receptor. As antigen concentration declines, less IL-2 is produced by T_H cells, and high-affinity IL-2 receptor levels decrease on antigen-specific CTLs and other effector T cells. However, if growth factor, but not antigen, is provided to the antigen-specific T cells, they neither proliferate nor die. It seems that IL-2 binding to the low-affinity receptor that is constitutively expressed on T cells may be sufficient to allow the cells to survive in the absence of antigenic stimulation. Under these in vitro conditions, IL-2 acts to inhibit apoptosis. Further investigation is required to understand the molecular basis for this observation.

Physiologically, apoptosis of antigen-responsive T cells upon antigen clearance is advantageous for at least two reasons. First, induction of programmed cell death upon antigen clearance assures rapid termination of an immune response without the metabolic costs involved in sustaining a large number of effector cells. Second, death of the expanded clones of IL-2-responsive effector cells prevents interference with subsequent immune responses to unrelated antigens by avoiding competition for limiting concentrations of growth factors. It appears that the immune system has evolved a complex set of controls that govern not only proliferation, but also death, of antigen-specific lymphocytes.

Apoptosis in Target Cells Induced by CTLs

Some of the IL-2-dependent effector T cells activated in response to infection are able to kill virally infected cells. These T cells are called killer or cytotoxic T lymphocytes. It is also possible to generate CTLs that kill tumor and/or allogeneic cells. We describe how CTLs kill, and we refer the reader to articles concerning apoptosis mediated by natural killer (NK) and lymphokine-activated killer (LAK) cells (Sanderson and Thomas 1977; Bishop and Whiting 1983; Stacey et al. 1985; Duke et al. 1986; Zychlinsky et al. 1991).

Characteristics of CTL-mediated Killing

CTLs bind to their "targets" (e.g., virally infected cells) via highly specific receptor-ligand interactions involving the T-cell antigen receptor and target-cell MHC molecules. Close contact between the CTL and target is required for lysis (Rosenau 1963; Wilson 1965). Because CTLs bind only to target cells bearing antigens for which the CTL is specific, bystander cells (e.g., noninfected cells within a tissue) are not killed, even when intimately mixed with appropriate targets (MacDonald et al. 1973; Cerottini and Brunner 1974; Gensheimer and Neefe 1978). The characteristics of CTL-mediated killing suggest that a simple, nonspecific, soluble mediator cannot be involved in the lytic process (Golstein and Smith 1977; Duke and Cohen 1988). How does the CTL exert its lethal effects?

The most popular models suggest that CTLs kill in a manner similar to that of complement. In these models, the CTL becomes triggered following binding to the target such that it secretes the contents of "lytic granules" into the intercellular space between the effector and target cell (Young and Cohn 1986). Lytic granules isolated from CTL clones contain a pore-forming protein, termed perforin or cytolysin (Henkart et al. 1984; Podack et al. 1985a,b), and enzymes including serine esterases (Pasternak and Eisen 1985; Masson et al. 1985). Perforin molecules insert into cell membranes, and in the presence of extracellular calcium, assemble into pore-like structures (Podack 1985). It has been postulated that once a threshold amount of membrane damage is attained, the target cell dies by colloid osmotic lysis, just as is observed during complement-mediated killing.

Whereas cells treated with complement or with purified lytic granules or perforin die by what seems to be necrosis, the targets of CTLs do not. The targets of CTLs appear to die via apoptosis (Cohen et al. 1985; Duvall and Wyllie 1986). This idea is based on observations of cell-mediated killing both in vitro and in vivo during viral infections, autoimmune reactions, and graft rejection (Battersby et al. 1974; Slavin and Woodruff 1974; Don et al. 1977; Sanderson and Glauert 1977; Searle et al. 1977; Matter 1979; Russell and Dobos 1980; Wyllie et al. 1980; Russell et al. 1982; Duke et al. 1983; Duke and Cohen 1988). By biochemical and morphological criteria, CTL-mediated killing is an example of programmed cell death (Table 2).

Nuclear damage in addition to DNA fragmentation can be detected in targets of CTLs. When target cells that contain fragmented DNA but that have not yet lysed are treated with nonionic detergents, the fragmented DNA does not sediment with the nucleus (Russell et al. 1980, 1982; Duke and Cohen 1988). This means that the DNA fragments are already in the

---

TABLE 2   *CHARACTERISTICS OF CTL-MEDIATED KILLING OF MOUSE CELLS*

1. Target cell nuclear damage, as determined either by chromatin condensation by light or electron microscopy; lack of sedimentation of chromatin following detergent lysis of cells in hypotonic buffer containing EDTA; or the appearance of oligonucleosome-sized fragments upon agarose gel electrophoresis of cellular DNA, occurs within *5–15 min* of interaction with the CTL.
2. Cell death, as measured by release of radioactive chromium [^{51}Cr], follows the induction of nuclear damage by *30–120 min.*
3. Nuclear damage and cell death occur *independently of de novo* protein synthesis.
4. Endogenous endonuclease activity *cannot* be detected in the nuclei of any of the target cells tested before or following interaction with the CTL.

Summary of results reported by Russell (1983), Duke et al. (1983), and Cohen et al. (1985). It should be noted that DNA cleavage also occurs in human target cells, although this damage often does not proceed to oligonucleosomes (Christiaansen and Sears 1985; Gromkowski et al. 1986; Howell and Martz 1987; Sellins et al. 1988).

cytoplasm or are still in the nucleus but are readily released from it by the detergent. This is a very different result from that obtained when DNA in isolated nuclei is cleaved with endogenous or added endonuclease: Almost all the fragmented DNA under these circumstances remains associated with the nuclei, even when treated with detergent (Duke and Cohen 1988). To release fragmented DNA from these nuclei requires treatment with trypsin or chelation of magnesium ions. These treatments disrupt overall chromatin superstructure. Thus, the appearance of detergent-soluble DNA after exposure to CTLs indicates that normal DNA-nucleoprotein interactions have been disrupted; that is, breakdown of overall nuclear structure occurs in addition to but not as a consequence of DNA fragmentation. In support of this idea, inhibitors of DNA topoisomerases or poly(ADP)ribosyl transferase block CTL-induced target cell DNA fragmentation and lysis (Duke and Sellins 1989; Nishioka and Welsh 1991). These enzymes are known to have dramatic effects on chromatin topology during transcription and replication (Liu et al. 1980; Park et al. 1983). It is reasonable to postulate that activation of both of these enzymes as part of a suicide program might account for the morphological changes observed in the target cell nucleus. In addition, altered chromatin could become a substrate for an endogenous endonuclease.

The similarities between programmed cell death and CTL-mediated cytolysis suggest that killer cells activate an endogenous suicide program in their targets; however, it is also clear that CTL-mediated killing is more complex than typical apoptosis. First, CTL-mediated DNA fragmentation and lysis occur in the presence of concentrations of protein synthesis inhibitors that completely abrogate programmed cell death (Duke et al. 1983). Second, cell death (lysis) occurs with faster kinetics following induction of DNA fragmentation in CTL-mediated killing than in the glucocorticoid, γ irradiation, or IL-2-deprivation systems. This is a real difference in mechanism, and not just due to differences in target cell phenotype, as is readily apparent when the kinetics of lysis of the same cells (CTLL-2) are compared during CTL-mediated cytolysis and following growth factor removal. CTLL-2 cells die approximately 60 minutes after induction of CTL-mediated

DNA fragmentation, as opposed to 4–6 hours after growth-factor-withdrawal-induced nuclear damage (Duke and Cohen 1988). Third, CTL-mediated killing often presents a mixed morphology of both necrotic and apoptotic cells (Zychlinsky et al. 1991). These observations suggest that CTLs use multiple cytolytic mechanisms to assure rapid killing (Young et al. 1987; Ostergaard and Clark 1989; Ju et al. 1990).

A Model of CTL-mediated Killing

Given the available data, the following model can be constructed to explain how CTLs kill (Fig. 1). Engagement of the CTL antigen receptor triggers the CTL to deliver the so-called lethal hit (Wagner and Rollinghoff 1974; Martz 1975). The lethal hit is operationally defined as the point from which the target cell is irreversibly "programmed for lysis" (Golstein and Smith 1977; Martz 1977). Programming for lysis is associated with secretion of lytic granule contents from the CTL (Henkart 1985; Pasternak and Eisen 1985; Podack 1985), a rise in intracellular calcium concentration in the target cell (Martz et al. 1983; Tsien and Poenie 1986), and induction of DNA fragmentation (Russell and Dobos 1980; Duke et al. 1983; Duke

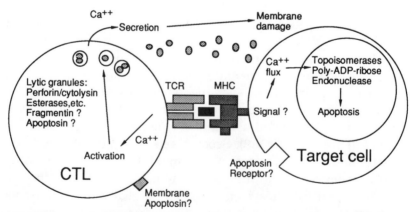

FIGURE 1 Model of CTL-mediated cytolysis. After binding to the target cell, the CTL is "triggered" such that (1) the CTL causes direct damage to the target cell's plasma membrane and (2) the CTL activates an intrinsic pathway in the target cell that leads to apoptotic-like nuclear damage.

and Cohen 1988). Thus, triggering of the CTL results in at least two things: (1) direct damage to the target cell membrane and (2) initiation of the process(es) that leads to induction of DNA cleavage and breakdown of overall nuclear structure.

The relative contributions of membrane and apoptotic-like internal damage are unknown at present, but experimental observations suggest the following associations: (1) Apoptotic-like nuclear damage synergizes with membrane damage leading to rapid lysis. (2) In the absence of lethal membrane damage, the target cells will lyse via apoptosis. (3) In the absence of induction of nuclear damage, lysis may not occur (target cell mutants that apparently lack the suicide program cannot be killed by CTLs [Ucker 1987], and zinc, a potent inhibitor of the endogenous endonuclease, blocks CTL-mediated DNA fragmentation as well as lysis [Duke et al. 1983; Cohen et al. 1985]) or may require higher numbers of killer cells bound to the target (Duke et al. 1986; Gromkowski et al. 1988).

How Does the CTL Induce Target Cell Apoptosis?

It is an intriguing idea that CTLs hold the secret to activating programmed cell death in any cell. Thus, knowing how the CTL induces apoptosis is of fundamental importance in understanding programmed cell death. After much research, there is no evidence to support any particular mechanism; however, two possibilities seem likely: (1) The CTL transfers the relevant enzymes or enzyme activators to the target cell or (2) the CTL transfers an activating factor or signals the target cell in such a way that an intrinsic cell death program involving nuclear damage is initiated.

Although the CTL could directly transfer the nuclear-modifying enzymes to the target, there is little evidence to support such an idea. For example, isolated lytic granules do not contain an endonuclease and cannot induce endonuclease activity in targets (Duke et al. 1988; Gromkowski et al. 1988). The pore-forming protein, perforin or cytolysin, is unable to induce the type of DNA fragmentation and apoptotic morphology observed during CTL-mediated killing, suggesting that mediators other than perforin are likely to be involved in the induction of nuclear damage (Duke et al. 1989). There have been several reports that certain preparations of lytic granules or their contents can induce apoptosis in cells (Konigsberg and Podack 1985; Munger et al. 1985, 1988; Hameed et al. 1989); however, these reports have not been verified (Duke et al. 1988, 1989, and in prep.; Gromkowski et al. 1988).

Tumor necrosis factor α (TNF-α) (Laster et al. 1988), TNF-β (Schmid et al. 1986), and a TNF-like molecule termed leukalexin (Liu et al. 1987) can induce apoptosis in tumor cells but are poor candidates for the mechanism of CTL-mediated DNA fragmentation. First, these molecules are cytotoxic for only a limited number of tumor cells, whereas CTLs can kill any normal (e.g., allogeneic), virally infected, or tumor cell to which it can bind. Second, TNF-related factors are slow-acting, inducing apoptosis in susceptible cells only after 24 hours or more of exposure. CTLs can induce DNA fragmentation in minutes. Last, although some CTLs express TNF-β and leukalexin, many do not (Ju et al. 1990). A soluble, rapidly acting, apoptosis-inducing factor cannot be ruled out, although the experiments describing such a molecule(s) (Munger et al. 1988; Duke et al. 1989) have been difficult to reproduce. However, a very recent report by Shi and colleagues (1991) describing a highly labile soluble factor, termed fragmentin, suggests that such a molecule may exist.

How the CTL would deliver an apoptosis-inducing factor is unclear but probably would not involve granule exocytosis. It has been shown that CTL-mediated killing can occur under conditions that prevent exocytosis (Ostergaard et al. 1987; Trenn et al. 1987, 1989; Takayama et al. 1988; Nash and Duke 1991). To prevent granule exocytosis, cyclosporin A, cholera toxin, or culture medium lacking calcium is employed. Under these conditions, CTLs induce lysis *and* DNA fragmentation (Ostergaard and Clark 1989). Thus, it seems most likely that the CTL might directly signal activation of a suicide program.

Returning to the model in Figure 1, the enzymes that induce DNA fragmentation and disrupt DNA-nucleoprotein interactions are already present in all cells as proenzymes. The proenzymes, when activated, encompass the final common pathway of programmed cell death. The CTL interacts with the target via specific target-cell membrane proteins including, but not limited to, MHC molecules. Recently, a membrane-associated factor(s) has been obtained from nonspecific LAK cells (Felgar and Hiserodt 1990). This factor induces apoptosis in any cell to which it is added and would certainly make an attractive candidate for an apoptosis-inducing, "skeleton-key" type, molecule. Perhaps the binding of this molecule, or the recently described soluble factor fragmentin (Shi et al. 1991), to some ubiquitously expressed receptor results in the transduction of a death signal to the target cell.

The evidence for such a signal transduction mechanism may be provided by the rise in target cell intracellular calcium concentration $[Ca^{++}]_i$, which temporally coincides with the target cell's being irreversibly programmed to die. Using Quin-2 or BAPTA to chelate target-cell intracellular free calcium prevents DNA fragmentation but does not always block killing (Hameed et al. 1989; McConkey et al. 1990). This observation is a reminder that the cytotoxic cell employs multiple cytolytic mechanisms to assure killing, but more importantly, it suggests that a rise in target cell intracellular calcium is required for the CTL to induce apoptosis. It should be noted that a rise in $[Ca^{++}]_i$ is not sufficient to induce apoptosis; a number of agents, including the calcium ionophores A23187 and ionomycin, as well as sublethal and lethal concentrations of the pore-forming proteins mellitin, staphylococcal α-toxin, complement, and perforin, do not mediate DNA fragmentation even though they all cause a detectable rise in intracellular calcium (R.C. Duke et al., in prep.). How the CTL induces apoptosis remains to be elucidated.

Conclusions: Questions to Be Answered about Apoptosis in Cell-mediated Immunity

Two examples of programmed cell death during cell-mediated immunity have been described. Several important questions with important implications for the general mechanism of apoptosis remain to be answered in these systems: (1) What are the proteins synthesized in IL-2-dependent cells upon removal of growth factor? (2) How do these "death" proteins lead to apoptosis? (3) What is the nature and regulation of the endonuclease that is activated during programmed cell death?

(4) What is the role for intracellular calcium in these systems? (5) Do cytotoxic T cells have a skeleton key to unlock cellular suicide?

Even without these answers, these systems and many more that are under investigation have provided invaluable insights into a phenomenon which is becoming more and more important to biologists; that phenomenon is apoptosis.

ACKNOWLEDGMENTS

The author is grateful to Drs. J. John Cohen and Karen Sellins and to Paul Nash for their many scholarly contributions to the ideas presented in this text. Our work was supported by grants from the U.S. Public Health Service of the National Institutes of Health (AI-11661 and AI-29953), the American Cancer Society, and the Pauline A. Morrison Charitable Trust.

REFERENCES

Battersby, C., W.S. Egerton, G. Balderson, J.F. Kerr, and W. Burnett. 1974. Another look at rejection of pig liver homografts. *Surgery* **76:** 617.

Bell, P.A. and C.N. Jones. 1982. Cytotoxic effects of butyrate and other "differentiation inducers" on immature lymphoid cells. *Biochem. Biophys. Res. Commun.* **104:** 1202.

Bishop, C.J. and V.A. Whiting. 1983. The role of natural killer cells in the intravascular death of intravenously injected murine tumor cells. *Br. J. Cancer* **48:** 441.

Bishop, C.J., D.J. Moss, J.M. Ryan, and S.R. Burrows. 1985. T lymphocytes in infectious mononucleosis. II. Response *in vitro* to interleukin-2 and establishment of T cell lines. *Clin. Exp. Immunol.* **60:** 70.

Bjorkman, P.J., M.A. Saper, B. Samoraui, W.S. Bennett, J.L. Strominger, and D.C. Wiley. 1987. The foreign antigen binding site and T cell recognition regions of class I histocompatibility antigens. *Nature* **329:** 512.

Buus, S., A. Sette, S. Colon, C. Miles, and H.M. Grey. 1985. The relation between major histocompatibility complex (MHC) restriction and the capacity of Ia to bind immunogenic peptides. *Science* **235:** 1353.

Cantrell, D.A. and K.A. Smith. 1984. The interleukin-2 T-cell system: A new cell growth model. *Science* **24:** 1312.

Cerottini, J.-C. and K.T. Brunner. 1974. Cell-mediated cytotoxicity, allograft rejection and tumor immunity. *Adv. Immunol.* **18:** 67.

Christiaansen, J.E. and D.W. Sears. 1985. Lack of lymphocyte-induced DNA fragmentation in human targets during lysis represents a species-specific difference between human and murine cells. *Proc. Natl. Acad. Sci.* **82:** 4482.

Cohen, J.J. 1991. Programmed cell death in the immune system. *Adv. Immunol.* (in press).

Cohen, J.J. and R.C. Duke. 1984. Glucocorticoid activation of a calcium-dependent endonuclease in thymocyte nuclei leads to cell death. *J. Immunol.* **132:** 38.

Cohen, J.J., V. Fadok, K.S. Sellins, and R.C. Duke. 1991. Apoptosis in the immune system. *Annu. Rev. Immunol.* (in press).

Cohen, J.J., R.C. Duke, R. Chervenak, K.S. Sellins, and L.K. Olson. 1985. DNA fragmentation in targets of CTL: An example of programmed cell death in the immune system. *Adv. Exp. Med. Biol.* **184:** 439.

Don, M.M., G. Ablett, C.J. Bishop, P.G. Bundesen, K.J. Searle, and J.F. Kerr. 1977. Death of cells by apoptosis following attachment of specifically allergized lymphocytes *in vitro*. *Aust. J. Exp. Biol. Med. Sci.* **55:** 407.

Duke, R.C. and J.J. Cohen. 1986. IL-2 addiction: Withdrawal of growth factor activates a suicide program in dependent T cells. *Lymphokine Res.* **5:** 289.

———. 1988. The role of nuclear damage in lysis of target cells by cytotoxic T lymphocytes. In *Cytolytic lymphocytes and complement: Effectors of the immune system* (ed. E.R. Podack), p. 35. CRC Press, Boca Raton, Florida.

Duke, R.C. and K.S. Sellins. 1989. Target cell nuclear damage in addition to DNA fragmentation during cytotoxic T lymphocyte-mediated killing. *Prog. Leukocyte Biol.* **9:** 311.

Duke, R.C., R. Chervenak, and J.J. Cohen. 1983. Endogenous endonuclease-induced DNA fragmentation: An early event in cell-mediated cytolysis. *Proc. Natl. Acad. Sci.* **80:** 6361.

Duke, R.C., J.J. Cohen, and R. Chervenak. 1986. Differences in target cell DNA fragmentation induced by mouse cytotoxic T lymphocytes and natural killer cells. *J. Immunol.* **137:** 1442.

Duke, R.C., K.S. Sellins, and J.J. Cohen. 1988. Cytotoxic lymphocyte-derived lytic granules do not induce DNA fragmentation in target cells. *J. Immunol.* **141:** 2191.

Duke, R.C., P.M. Persechini, S. Chang, C.-C. Liu, J.J. Cohen, and J.D.-E. Young. 1989. Purified perforin induces target cell lysis but not DNA fragmentation. *J. Exp. Med.* **170:** 1451.

Duvall, E. and A.H. Wyllie. 1986. Death and the cell. *Immunol. Today* **7:** 115.

Felgar, R. and J. Hiserodt. 1990. A novel cytotoxic activity isolated from the plasma membrane of highly purified lymphokine activated killer cells and the LGL tumor CRNK-16. *FASEB J.* **4:** A1900.

Gensheimer, G.G. and J.R. Neefe. 1978. Cell mediated lympholysis: A receptor associated lytic mechanism. *Cell. Immunol.* **36:** 54.

Gillis, S., M.M. Ferm, W. Ou, and K.A. Smith. 1978. T cell growth factor: Parameters of production and a quantitative microassay for activity. *J. Immunol.* **120:** 2027.

Golstein, P. and E.T. Smith. 1977. Mechanism of T-cell-mediated cytolysis: The lethal hit stage. *Contemp. Top. Immunobiol.* **7:** 273.

Gromkowski, S.H., T.C. Brown, P.A. Cerutti, and J.-C. Cerottini. 1986. DNA of human Raji cells is damaged upon lymphocyte-mediated lysis. *J. Immunol.* **136:** 752.

Gromkowski, S.H., T.C. Brown, D. Masson, and J. Tschopp. 1988. Lack of DNA degradation in target cells lysed by granules derived from cytolytic T lymphocytes. *J. Immunol.* **141:** 774.

Hameed, A., K.J. Olsen, M.-K. Lee, M.G. Lichtenheld, and E.R. Podack. 1989. Cytolysis by Ca-permeable transmembrane channels. Pore formation causes extensive DNA degradation and cell lysis. *J. Exp. Med.* **169:** 765.

Henkart, P.A. 1985. Mechanism of lymphocyte-mediated cytotoxicity. *Annu. Rev. Immunol.* **3:** 31.

Henkart, P.A., P.J. Millard, C.W. Reynolds, and M.P. Henkart. 1984. Cytolytic activity of purified cytoplasmic granules from cytotoxic rat LGL tumors. *J. Exp. Med.* **160:** 75.

Howell, D.M. and E. Martz. 1987. The degree of CTL-induced DNA solubilization is not determined by the human vs. mouse origin of the target cell. *J. Immunol.* **138:** 3695.

Ju, S.-T., N.H. Ruddle, P. Struck, M.E. Dorf, and R.H. DeKruyff. 1990. Expression of two distinct cytolytic mechanisms among murine CD4 subsets. *J. Immunol.* **144:** 23.

Konigsberg, P.J. and E.R. Podack. 1985. Target cell DNA fragmentation induced by cytolytic T-cell granules. *J. Leukocyte Biol.* **38:** 109.

Laster, S.M., J.G. Wood, and L.R. Gooding. 1988. Tumor necrosis factor can induce both apoptotic and necrotic forms of cell lysis. *J. Immunol.* **141:** 2629.

Liu, C.-C., M. Steffen, F. King, and J. D.-E. Young. 1987. Identification, isolation, and characterization of a novel cytotoxin in murine cytolytic lymphocytes. *Cell* **51:** 393.

Liu, L.F., C.-C. Liu, and B.M. Alberts. 1980. Type II DNA topoisomerases: Enzymes that can unknot a topologically knotted DNA molecule via a reversible double-stranded break. *Cell* **19:** 697.

MacDonald, H.R., R.A. Phillips, and R.G. Miller. 1973. Allograft immunity in the mouse. I. Quantitation and specificity of cytotoxic effector cells after *in vitro* sensitization. *J. Immunol.* **111:** 565.

Martz, E. 1975. Early steps in specific tumor cell lysis by sensitized mouse T-lymphocytes. I. Resolution and characterization. *J. Immunol.* **115:** 261.

―――. 1977. Mechanism of specific tumor cell lysis by alloimmune T lymphocytes: Resolution and characterization of discrete steps is in the cellular interaction. *Contemp. Top. Immunobiol.* **7:** 301.

Martz, E., W. Heagy, and H. Gromkowski. 1983. The mechanism of CTL-mediated killing: Monoclonal antibody analysis of the roles of killer and target cell membrane proteins. *Immunol. Rev.* **72:** 73.

Masson, D., P. Corthesy, M. Nabholz, and J. Tschopp. 1985. Appearance of cytolytic granules upon induction of cytolytic activity in CTL-hybrids. *EMBO J.* **4:** 2533.

Matter, A. 1979. Microcinematographic and electron microscopic analysis of target cell lysis induced by cytotoxic T lymphocytes. *Immunology* **36:** 179.

McConkey, D.J., S.C. Chow, S. Orrenius, and M. Jondal. 1990. NK cell-induced cytotoxicity is dependent on a Ca^{2+} increase in the target. *FASEB J.* **4:** 2661.

McConkey, D.J., P. Hartzell, M. Jondal, and S. Orrenius. 1989. Calcium-dependent killing of immature thymocytes by stimulation via the CD3/T cell receptor complex. *J. Immunol.* **143:** 1801.

McConkey, D.J., P. Hartzell, S.K. Duddy, H. Hakansson, and S. Orrenius. 1988. 2,3,7,8-tetrachlorodibenzo-*p*-dioxin kills immature thymocytes by Ca^{2+} mediated endonuclease activation. *Science* **242:** 256.

Munger, W.E., G.A. Berrebi, and P.A. Henkart. 1988. Possible involvement of CTL granule proteases in target cell DNA breakdown. *Immunol. Rev.* **103:** 99.

Munger, W.E., C.W. Reynolds, and P.A. Henkart. 1985. DNase activity in cytoplasmic granules of cytotoxic lymphocytes. *Fed. Proc.* **44:** 1284.

Nash, P.B. and R.C. Duke, 1991. Certain requirements for delivery of the lethal hit can be overcome by "triggering" the effector cells prior to interaction with their targets. *FASEB J.* **5:** A600.

Nishioka, W.K. and R.M. Welsh. 1991. Inhibition of cytotoxic T lymphocyte-induced target cell DNA fragmentation, but not lysis, by inhibitors of DNA topoisomerases I and II. *FASEB J.* **5:** A601.

Ostergaard, H.L. and W.R. Clark. 1989. Evidence for multiple lytic pathways used by cytotoxic T lymphocytes. *J. Immunol.* **143:** 2120.

Ostergaard, H.L., K.P. Kane, M.F. Mescher, and W.R. Clark. 1987. Cytotoxic T lymphocyte mediated lysis without release of serine esterase. *Nature* **330:** 71.

Park, S.D., C.G. Kim, and M.G. Kim. 1983. Inhibitors of poly (ADP-ribose) polymerase enhance DNA strand breaks, excision repair and sister chromatid exchanges induced by alkylating agents. *Environ. Mutagen.* **5:** 515.

Pasternak, M.S. and H.N. Eisen. 1985. A novel serine esterase expressed by cytotoxic T lymphocytes. *Nature* **314:** 743.

Podack, E.R. 1985. The molecular mechanism of lymphocyte-mediated tumor cell lysis. *Immunol. Today* **6:** 21.

Podack, E.R., J.D.-E. Young, and Z.A. Cohn. 1985a. Isolation and biochemical and functional characterization of perforin 1 from cytolytic T cell granules. *Proc. Natl. Acad. Sci.* **82:** 8629.

Podack, E.R., P.J. Konigsberg, H. Acha-Orbea, H. Pircher, and H. Hengartner. 1985b. Cytolytic T-cell granules: Biochemical properties and functional specificity. *Adv. Exp. Med. Biol.* **184:** 99.

Robb, R.J., C.M. Rusk, J. Yodoi, and W.C. Greene. 1987. Interleukin 2 binding molecule distinct from the Tac protein: Analysis of its role in formation of high affinity receptors. *Proc. Natl. Acad. Sci.* **84:** 2002.

Rosenau, W. 1963. Interaction of lymphoid cells with target cells in tissue culture. In *Cell-bound antibodies* (ed. B. Amos and H. Koprowski), p. 75. The Wistar Institute Press, Philadelphia.

Russell, J.H. 1983. Internal disintegration model of cytotoxic lymphocyte-induced target damage. *Immunol. Rev.* **72:** 97.

Russell, J.H. and C.B. Dobos. 1980. Mechanisms of immune lysis. II. CTL-induced nuclear disintegration of the target begins within minutes of cell contact. *J. Immunol.* **125:** 1256.

Russell, J.H., V.R. Masakowski, and C.B. Dobos. 1980. Mechanisms of immune lysis. I. Physiological distinction between target cell death mediated by cytotoxic T lymphocytes and antibody plus complement. *J. Immunol.* **124:** 1100.

Russell, J.H., V. Masakowski, T. Rucinsky, and G. Phillips. 1982. Mechanisms of immune lysis. III. Characterization of the nature and kinetics of the cytotoxic T lymphocyte-induced nuclear lesion in the target. *J. Immunol.* **128:** 2087.

Sanderson, C.J. and A.M. Glauert. 1977. The mechanism of T-cell mediated cytolysis. V. Morphological studies by electron microscopy. *Proc. R. Soc. Lond. B Biol. Sci.* **198:** 315.

Sanderson, C.J. and J.A. Thomas. 1977. The mechanism of T-cell mediated cytolysis. II. Characteristics of the effector cell and morphological changes in the target cell. *Proc. R. Soc. Lond. B Biol. Sci.* **197:** 417.

Schmid, D.S., J.P. Tite, and N.H. Ruddle. 1986. DNA fragmentation: A manifestation of target cell destruction mediated by cytotoxic T cell lines, lymphotoxin-secreting helper T cell clones, and cell-free lymphotoxin containing supernatant fluids. *Proc. Natl. Acad. Sci.* **83:** 1881.

Searle, J., J.F.R. Kerr, C. Battersby, W.S. Egerton, G. Balderson, and W. Burnett. 1977. An electron microscopic study of the mode of donor cell death in unmodified rejection of pig liver allografts. *Aust. J. Exp. Biol. Med. Sci.* **55:** 401.

Sellins, K.S. and J.J. Cohen. 1987. Gene induction by gamma-irradiation leads to DNA fragmentation in lymphocytes. *J. Immunol.* **139:** 3199.

————. 1991. Hyperthermia induces apoptosis in thymocytes. *Radiat. Res.* **126:** 88.

Sellins, K.S., R.C. Duke, and J.J. Cohen. 1988. All target cells undergo DNA damage during cytotoxic T cell-mediated lysis. *FASEB J.* **2:** A468.

Sharon, M., R.D. Klausner, B.R. Cullen, R. Chizzonite, and W.J. Leonard. 1986. Novel interleukin 2 receptor subunit detected by cross-linking under high-affinity conditions. *Science* **234:** 859.

Shi, Y.F., R.P. Kraut, and A.H. Greenberg. 1991. Partial purification of NK cell granule nucleolytic factor. *FASEB J.* **5:** A601.

Shi, Y.F., B.M. Sahai, and D.R. Green. 1989. Cyclosporin A inhibits activation-induced cell death in T-cell hybridomas and thymocytes. *Nature* **339:** 625.

Skalka, R.E., J. Mattyasova, and M. Cejkova. 1976. DNA in chromatin of irradiated lymphoid tissues degrades *in vivo* into regular fragments. *FEBS Lett.* **72:** 271.

Slavin, R.E. and J.M. Woodruff, J.M. 1974. The pathology of bone marrow transplantation. *Pathol. Annu.* **9:** 291.

Smith, C.A., G.T. Williams, R. Kingston, E.J. Jenkinson, and J.J.T. Owen. 1989. Antibodies to CD3/T-cell receptor complex induce death by apoptosis in immature T cells in thymic cultures. *Nature* **337:** 181.

Stacey, N.H., C.J. Bishop, J.W. Halliday, W.J. Halliday, W.G.E. Cooksley, L.W. Powell, and J.F.R. Kerr. 1985. Apoptosis as the mode of cell death in antibody-dependent lymphocytotoxicity. *J. Cell Sci.* **74:** 169.

Takayama, H., G. Trenn, and M.V. Sitkovsky. 1988. Locus of inhibitory action of cAMP-dependent protein kinase in the antigen receptor-triggered cytotoxic T lymphocyte activation pathway. *J. Biol. Chem.* **263:** 2330.

Teshigawara, K., H.M. Wang, K. Kato, and K.A. Smith. 1987. Interleukin 2 high-affinity receptor expression requires two distinct binding proteins. *J. Exp. Med.* **165:** 223.

Trenn, G., H. Takayama, and M.V. Sitkowski. 1987. Exocytosis of cytolytic granules may not be required for target cell lysis by cytotoxic T-lymphocytes. *Nature* **330:** 72.

Trenn, G., R. Taffs, R. Hohman, R. Kincaid, E.M. Shevach, and M.V. Sitkovsky. 1989. Biochemical characterization of the inhibitory effect of CsA on cytolytic T lymphocyte effector functions. *J. Immunol.* **142:** 3796.

Tsien, R.Y. and M. Poenie. 1986. Fluorescence ratio imaging: A new window into intracellular ionic signaling. *Trends Biochem. Sci.* **11:** 450.

Ucker, D.S. 1987. Cytotoxic T lymphocytes and glucocorticoids activate an endogenous suicide process in target cells. *Nature* **327:** 62.

Vanderbilt, J.N., K.S. Bloom, and J.N. Anderson. 1982. Endogenous nuclease. Properties and effects on transcribed genes in chromatin. *J. Biol. Chem.* **257:** 13009.

Wagner, H. and M. Rollinghoff. 1974. T cell mediated cytotoxicity: Discrimination between antigen recognition, lethal hit and cytolysis phase. *Eur. J. Immunol.* **4**: 745.

Watts, T.H., J. Gariepy, G.K. Schoolnik, and H.M. McConnell. 1985. T-cell activation by peptide antigen: Effect of peptide sequence and method of antigen presentation. *Proc. Natl. Acad. Sci.* **82**: 5480.

Weissman, A.M., J.B. Harford, P.B. Svetlik, W.J. Leonard, J.M. Depper, T.A. Waldmann, W.C. Greene, and R.D. Klausner. 1986. Only high affinity receptors for interleukin-2 mediate internalization of ligand. *Proc. Natl. Acad. Sci.* **83**: 146.

Wilson, D. 1965. Quantitative studies on the behavior of sensitized lymphocytes in vitro. I. Relationship of the degree of destruction of homologous target cells to the number of lymphocytes and to the time of contact in culture and consideration of iso-immune serum. *J. Exp. Med.* **122**: 143.

Wyllie, A.H. 1980. Glucocorticoid-induced thymocyte apoptosis is associated with endogenous endonuclease activation. *Nature* **284**: 555.

Wyllie, A.H., J.F.R. Kerr, and A.R. Currie. 1980. Cell death: The significance of apoptosis. *Int. Rev. Cytol.* **68**: 251.

Young, J.D.-E. and Z.A. Cohn. 1986. Cell-mediated killing: A common mechanism? *Cell* **46**: 641.

Young, J.D.-E., W.R. Clark, C.-C. Liu, and Z.A. Cohn. 1987. A calcium- and perforin-independent pathway of killing mediated by murine cytolytic lymphocytes. *J. Exp. Med.* **160**: 1894.

Zinkernagel, R.M. and P.C. Doherty. 1974. Activity of sensitized thymus derived lymphocytes in lymphocytic choriomeningitis reflects immunological surveillance against altered self components. *Nature* **251**: 547.

Zvonareva, N.B., B.D. Zhitovsky, and K.P. Hanson. 1983. Distribution of nuclease attack sites and complexity of DNA in the products of post-irradiation degradation of rat thymus chromatin. *Int. J. Radiat. Biol.* **44**: 261.

Zychlinsky, A., L.M. Zheng, C.-C. Liu, and J.D.-E. Young. 1991. Cytolytic lymphocytes induce both apoptosis and necrosis in target cells. *J. Immunol.* **146**: 393.

Cellular Signaling in Thymocyte Apoptosis

D.J. McConkey[1] and S. Orrenius[2]

[1]Laboratory of Immunobiology, Dana-Farber Cancer Institute
and Department of Pathology, Harvard Medical School
Boston, Massachusetts 02115
[2]Department of Toxicology, Karolinska Institutet
S-104 01 Stockholm, Sweden

Timed induction of apoptosis appears to play an important role in normal cell turnover and development (Wyllie et al. 1980; Lockshin 1981). For example, up to 90% of immature thymocytes die by apoptosis in situ, and increasing evidence suggests that this response is a critically important component of the intrathymic cell selection process that functions to prevent autoimmunity. Therefore, it appears likely that physiological mechanisms control apoptosis in vivo.

Ongoing work has identified several biochemical mechanisms that appear to regulate apoptosis. Much of this information has been gained from studies on immature thymocytes, which undergo apoptosis in response to a number of stimuli (Table 1). The striking feature of these pathways is that they are generally similar or identical to some of the relatively well-defined signaling processes regulating cell division. In this paper, we review what is known about the signaling processes regulating apoptosis in immature thymocytes within the context of what is known about lymphoid cell proliferation, thymic cell selection, and the development of tolerance to self-antigens. The general relevance of these pathways to apoptosis in other cell types is discussed.

Cellular Signaling during T-lymphocyte Proliferation

An important part of the immune response is the clonal expansion of immunoreactive B and T lymphocytes. This is ac-

Apoptosis: The Molecular Basis of Cell Death
Copyright 1991 Cold Spring Harbor Laboratory Press 0-87969-366-5/91 $3.00 + 00

TABLE 1 *STIMULI KNOWN TO INDUCE THYMOCYTE APOPTOSIS*

Stimulus	Possible mediator	Reference
Glucocorticoids	Ca^{++}	Wyllie (1980)
Ca^{++} ionophores	Ca^{++}	Cohen and Duke (1984)
TCR stimulation	Ca^{++}	Smith et al. (1989)
Irradiation	poly(ADP-ribosylation)	Nelipovich et al. (1988)
Adenosine	poly(ADP-ribosylation)	Kizaki et al. (1988)
PGE_2, cAMP	cAMP	McConkey et al. (1990c)
Dioxin	Ca^{++}	McConkey et al. (1988)

complished in large part via specific surface structures known as antigen receptors that selectively bind to particular foreign antigens. On B lymphocytes, surface-expressed immuno-globulins serve as the major antigen-binding component of the B-cell antigen receptor, whereas T lymphocytes express a distinct but related dimeric structure, termed Ti, as the antigen-binding component of the T-cell receptor (TCR). Stimulation of B or T lymphocytes via their specific antigen receptors is a requirement for clonal expansion and immune effector functions.

The TCR is a surface complex consisting of at least seven chains, detailed in Figure 1. Heterogeneity is generated by gene recombination events in the variable regions of Ti α and β genes (Fowlkes and Pardoll 1989) in a fashion analogous to the generation of antibody diversity (Tonegawa 1983). The Ti chains are expressed on the surface of T cells in association with five invariant chains, termed CD3. It is thought that the CD3 subunits direct intracellular signal transduction mechanisms.

Arguably, most of the information on cellular proliferative signaling available to date has come from studies with T lymphocytes; these processes are now known to mediate the activation of other cell types as well. Specific anti-Ti or anti-CD3 antibodies induce T-cell proliferation by stimulating the synthesis of the autocrine growth factor, interleukin 2 (IL-2), which then promotes transition into the S phase of the cell cycle (Meuer et al. 1984). Several second-messenger systems have been implicated in TCR-mediated IL-2 production (Fig. 1). The initial event appears to involve activation of tyrosine ki-

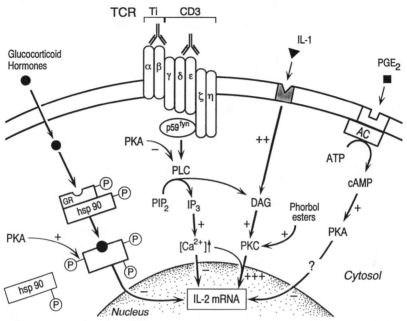

FIGURE 1 Schematic summary of the signaling processes regulating proliferative homeostasis in T lymphocytes. (GR) Glucocorticoid receptor; (PLC) phospholipase C; (DAG) diacylglycerol; (IP3) inositol-1,4,5-triphosphate; (PKC) protein kinase C; (AC) adenylate cyclase; (PKA) cAMP-dependent protein kinase; (IL-2) interleukin 2.

nase(s), resulting in the phosphorylation of a number of relevant cytoplasmic substrates. Tyrosine phosphorylation somehow mediates the activation of phospholipase C (PLC) (Mustelin et al. 1990), which hydrolyzes phosphatidylinositol 4,5-biphosphate, leading to the production of at least two critical second messengers, inositol triphosphate and diacylglycerol (Berridge 1984). The former binds to specific receptors on the endoplasmic reticulum to facilitate Ca^{++} release (Berridge and Irvine 1984; Ross et al. 1989), generating a transient increase in the cytosolic Ca^{++} concentration, and the latter stimulates protein kinase C (PKC) (Nishizuka 1984). TCR stimulation also leads to Ca^{++} influx through receptor-operated Ca^{++} channels in the plasma membrane, leading to a sustained Ca^{++} increase (Imboden and Weiss 1987). Accessory cell-derived growth factor, in particular interleukin 1 (IL-1), also appears to be required for cell division (Crabtree 1989), perhaps by augmenting diacylglycerol production (Rosoff et al. 1988).

Several observations support the notion that an elevation of the cytosolic Ca^{++} concentration and PKC activation are sufficient to promote T-cell proliferation. Bypassing TCR signaling by incubating T cells with Ca^{++} ionophores plus phorbol esters, tumor promoters known to stimulate PKC (Nishizuka 1984), results in proliferation (Truneh et al. 1985). In addition, analysis of the regulatory elements upstream of the IL-2 gene has revealed enhancer elements responding preferentially to Ca^{++} or PKC, and their simultaneous occupancy leads to optimal transcriptional activity (Crabtree 1989).

Several suppressor mechanisms have also been identified that block TCR-mediated T-cell proliferation (Fig. 1). Prostaglandins and certain pharmacological agents inhibit TCR-mediated second-messenger production and IL-2 gene induction by increasing the cAMP level (Kammer 1988). Glucocorticoid hormones inhibit proliferation in mature T cells by a mechanism that also involves inhibition of IL-2 production (Goodwin et al. 1986). Even certain examples of TCR triggering in the absence of accessory cell growth factor production can result in a state of prolonged proliferative nonresponsiveness, due to a defect in IL-2 production (Jenkins et al. 1987; Rammensee et al. 1989). This state is dependent on a sustained Ca^{++} increase and can be induced by Ca^{++} ionophore (Jenkins et al. 1987). It has been proposed that these suppressor mechanisms are important for maintaining tolerance to certain antigens, including those constitutively expressed to immune cells by host tissues (Schwartz 1990).

Development and Education of T Lymphocytes in the Thymus

As noted above, immune diversity among T cells is generated by Ti gene recombination in developing immature cells. As is true for antibody diversification, the process is essentially random, and as a result, many TCRs arise that would exhibit binding and reactivity to host tissue antigens in the periphery. Protective mechanisms have therefore evolved to silence or eliminate this "self-reactivity" to prevent deleterious autoimmunity. The overall state of self-nonreactivity is known as tolerance and appears to involve many different biochemical processes. These systems sometimes fail, as evidenced by the recent confirmation that host T cells play a predominant role in the pathology of the autoimmune diseases multiple sclerosis (Leiter 1989) and insulin-dependent diabetes mellitus (Calder et al. 1989).

Mature T lymphocytes do not develop in the absence of a thymus, and it is within this tissue compartment that developing T cells are "educated." Clones recognizing foreign antigens presented in the context of self major histocompatibility (MHC) gene products are positively selected and emigrate from the thymus to the pool of mature T lymphocytes in the periphery, whereas clones that react with self-antigens are destroyed (Blackman et al. 1990). Because the TCR is involved in antigen recognition, it is not surprising that it mediates both positive (McDuffie et al. 1986) and negative (White et al. 1989) selection in the thymus. What appears paradoxical is how one receptor structure (TCR) expressed on a given type of cell can induce essentially opposite responses. One model commonly evoked is based on the affinity of the TCR for antigen: Self-reactive cells are deleted by "strong" signaling, whereas more "moderate" self-interactions lead to positive selection based solely on self-MHC recognition (Blackman et al. 1990). Another hypothesis is based on the types of accessory cells found within the subcompartments of the thymus. In this model, both positive and negative selection involve high-affinity TCR interactions, but certain accessory cells are capable of delivering the additional signal(s) required for proliferation, whereas others are incapable, thereby inducing deletion (McConkey et al. 1990b). Finally, different TCR isotypes may deliver qualitatively different signals to cells undergoing positive or negative selection, as has been suggested by the work of Ashwell and colleagues (Mercep et al. 1988, 1989). These models are not mutually exclusive.

Little is known about the specific mechanisms underlying positive T-cell selection, except that the TCR and two surface molecules, CD4 (Zuniga-Pflucker et al. 1989) and CD8 (Zuniga-Pflucker et al. 1990), are required. The physiological role of the latter appears to be to recognize invariant determinants on MHC class II and class I molecules, respectively, which may explain why CD4 is found on T cells that react to MHC class II and CD8 on those responding to class I. Relatively more information is available on the events involved in negative selection and nondeletional tolerance induction in the thymus. Antibodies to the TCR stimulate apoptosis specifically in the population of transitional $CD4^+CD8^+$ immature thymocytes known to be undergoing both positive and negative selection (McConkey et al. 1989c; Shi et al. 1989; Smith et al. 1989). An identical response is observed in transgenic mice that express a particular TCR following treatment with a specific antigenic peptide in vivo (Murphy et al. 1990). These observations have led to the proposal that TCR-triggered apoptosis mediates negative selection in the thymus.

Cellular Signaling in Thymocyte Apoptosis

Role of Ca^{++}. Several independent observations have supported a critical role for the calcium ion in thymocyte apoptosis. Important early observations by Kaiser and Edelman showed that the cytolytic effects of glucocorticoids involve Ca^{++} influx (Kaiser and Edelman 1977) and can be mimicked by Ca^{++} ionophores (Kaiser and Edelman 1978). Cohen and Duke (1984) later described a nuclear, Ca^{++}-dependent and Zn^{++}-sensitive endonuclease whose activity appears to be responsible for the chromatin cleavage observed in glucocorticoid-treated thymocytes. In addition, Wyllie and co-workers (Wyllie et al. 1984) showed that Ca^{++} ionophore induces both the DNA fragmentation and morphological alterations typical of glucocorticoid-induced thymocyte apoptosis. Notably, they also showed that the responses to Ca^{++} ionophore treatment are blocked by cycloheximide or actinomycin D, effects they linked directly to reduced levels of protein and mRNA synthesis, respectively.

We designed experiments to directly investigate the role of Ca^{++} in thymocyte apoptosis. In glucocorticoid-treated thymocytes, an early, sustained Ca^{++} increase occurs before DNA fragmentation or loss of cell viability can be detected; intracellular Ca^{++} buffering or chelation of extracellular Ca^{++} blocks both endonuclease activation and cell death (McConkey et al. 1989d). An analogous mechanism is observed in thymo-cytes exposed to the toxic agent, 2,3,7,8-tetrachlorodibenzo-*p*-dioxin (TCDD) (McConkey et al. 1988). Up to a point, a tight dose-response relationship between the cytosolic Ca^{++} level and the DNA fragmentation is observed in thymocytes treated with Ca^{++} ionophore (McConkey et al. 1989b), suggesting that the cytosolic Ca^{++} increase may be rate-limiting for endonuclease activation. Because DNA fragmentation may be directly involved in triggering cell death in apoptotic thymocytes (McConkey et al. 1989b) and murine T-T hybrid-omas (Shi et al. 1990), these observations support the idea that Ca^{++} plays a critical role in cell death by apoptosis.

We also investigated the role of Ca^{++} in TCR-mediated apop-tosis in human thymocytes (McConkey et al. 1989c). Incuba-tion of the cells with anti-CD3 antibody resulted in a sustained Ca^{++} increase that preceded endonuclease activa-tion and cell death. In the presence of intracellular or ex-tracellular Ca^{++} chelators, both DNA fragmentation and cell death were blocked. Isolated peripheral blood T lymphocytes did not undergo apoptosis in response to TCR stimulation, despite the fact that a sustained Ca^{++} increase was also ob-served in the cells. Moreover, although Ca^{++} ionophore treat-ment resulted in apoptosis in the immature thymocytes, ma-ture T cells were completely resistant to ionophore treatment for at least 24 hours. Interestingly, glucocorticoid, Ca^{++} ionophore, and anti-CD3 antibody treatment resulted in the death of the same $CD4^{+}CD8^{+}$ thymocyte subpopulation. This selectivity for immature thymocytes suggests that suscep-tibility to apoptosis is developmentally regulated, because Ca^{++} ionophore induces identical Ca^{++} increases in immature thymocytes and mature T cells. We speculate that the acquisi-tion of sensitivity to apoptosis is likely to be an important determinant in intrathymic T-cell development; a hypothetical model for how this sensitivity could be induced will be dis-cussed later.

Role of PKC. Many investigators have observed that T lymphocytes do not proliferate when stimulated with anti-TCR antibodies alone but require phorbol esters or growth factors as well. This conclusion has led to the "two-signal hypothesis" of T-lymphocyte activation. Cellular and genetic work discussed above suggests that these second signals involve PKC activation (Crabtree 1989). Moreover, TCR stimulation in the absence of second signals can be inhibitory, leading to a state of proliferative nonresponsiveness, or anergy (Schwartz 1990). It has been proposed that in the periphery, clonal anergy may contribute to antigen nonresponsiveness, or tolerance (Rammensee et al. 1989).

On the basis of our findings on the role of Ca^{++}, we hypothesized that, in addition to mediating peripheral tolerance, unbalanced signaling could also underlie thymocyte apoptosis. To investigate this possibility, we tested the effects of phorbol esters on Ca^{++} ionophore-induced thymocyte apoptosis. We found these compounds to be good inhibitors of both DNA fragmentation and cell death in thymocytes without detectably affecting the cytosolic Ca^{++} level (McConkey et al. 1989a). We also found that phorbol ester treatment blocked both DNA fragmentation and cell death in response to anti-CD3 stimulation in immature $CD4^+CD8^+$ thymocytes (McConkey et al. 1989c). Moreover, phorbol ester treatment directly reduced Ca^{++}-dependent endonuclease activity in isolated thymocyte nuclei (McConkey et al. 1989a, 1990a). This reduction in endogenous nuclear endonuclease activity is also observed following treatment of thymocytes with cycloheximide in vitro (McConkey et al. 1990a) or in vivo (Umansky et al. 1981).

Cells do not naturally encounter phorbol esters. Therefore, to determine whether the results obtained with phorbol esters had any physiological relevance, we tested the effects of known T-cell growth factors on thymocyte apoptosis. IL-1 was originally identified by its capacity to support thymocyte proliferation in the presence of lectin mitogens (Durum et al. 1985), and its effects on T cells have been linked to the production of diacylglycerol (Rosoff et al. 1988). We found IL-1 to be an effective inhibitor of anti-CD3-induced thymocyte apoptosis (McConkey et al. 1990e). To our surprise, IL-2 was not effective; this may be related to an inherent lack of responsiveness to IL-2 in immature thymocytes (Boyer and Rothenberg 1989). Supporting previous work in mature T cells, IL-1 induced the rapid accumulation of diacylglycerol in thymocytes, suggesting that the effects may have been mediated by PKC.

Incubation of thymocytes with IL-1 also led to a loss of endonuclease activity in isolated thymocyte nuclei (McConkey et al. 1990a).

Role of cAMP. Agents that elevate cAMP inhibit T-cell activation (Kammer 1988) (Fig. 1), and it has been proposed that cAMP may be involved in certain pathways of T-cell tolerance induction (Johnson et al. 1988). In addition, previous work has linked cAMP to programmed cell death during palatal development (Pratt and Martin 1975) and in some plants (Basile et al. 1973). Previous work had linked cAMP to the killing of certain sensitive lymphoid cell lines (Daniel et al. 1973) and thymocytes (Kristensen et al. 1983), although no mechanism of action was defined. We therefore questioned whether cAMP might also regulate endonuclease activation in thymocytes.

E-series prostaglandins (PGEs) are macrophage and dendritic cell products that are potent inhibitors of T-cell activation, and their inhibitory actions have been linked to accumulation of cAMP (Kammer 1988). Treatment of rat thymocytes with PGE_2 or pharmacological agents that directly raise cAMP levels resulted in rapid DNA fragmentation characteristic of apoptosis (McConkey et al. 1990c). The mechanism was linked to the activation of protein kinase A (PKA). Preincubation of the thymocytes with phorbol esters or IL-1 inhibited the response, suggesting "cross-talk" between the PKA and PKC signaling pathways. We did not detect a cAMP-stimulated increase in the cytosolic Ca^{++} concentration, although intracellular Ca^{++} buffering blocked cAMP-induced DNA fragmentation and cell death. Therefore, cAMP-induced thymocyte apoptosis appears to be dependent on Ca^{++}, although how this dependency is expressed is unclear at present.

We also wondered whether cAMP could influence the other pathways of apoptosis in thymocytes. Although we found that cAMP had no effect on Ca^{++} ionophore-induced DNA fragmentation, cAMP dramatically potentiated glucocorticoid-induced cytosolic Ca^{++} increases and DNA fragmentation in murine thymocytes (D.J. McConkey et al., in prep.). Glucocorticoid receptor (GR) binding was enhanced by cAMP, consistent with previous results (Gruol et al. 1986), and immunoprecipitation of the GR from cAMP- and glucocorticoid-treated thymocytes revealed significant potentiation of GR phosphorylation by cAMP. Thus, cAMP likely promotes GR function via PKA activation and subsequent GR phosphorylation. It is likely that cAMP plays a general role in the regulation of steroid receptor function, as suggested by the results of O'Malley and colleagues, who showed that cAMP potentiates progesterone receptor-mediated transcriptional activation (Denner et al. 1990).

Role of poly(ADP-ribose) polymerase. Gamma irradiation is another means of inducing apoptosis in thymocytes (Afanasyev et al. 1986; Sellins and Cohen 1987; Nelipovich et al. 1988). Because a primary lesion associated with ionizing irradiation is DNA single-strand breaks, researchers have sought mechanisms relevant to DNA repair to explain the response. This work has implicated poly(ADP-ribose) polymerase, a DNA repair enzyme (Lunee 1984) that is also involved in the regulation of lymphoid cell differentiation (Johnstone et al. 1982). DNA strand breaks directly stimulate the enzyme in vitro, with double-strand cleavage exhibiting more potency than single-strand breaks (Benjamin and Gill 1980). Using NAD$^+$ as a substrate, the enzyme links poly(ADP-ribose) moieties to cleaved sites in the DNA backbone, presumably allowing for the short-term maintenance of structural integrity until DNA polymerases can make permanent repairs. Nuclear proteins, including histones and the putative Ca^{++}-dependent endonuclease, can also serve as acceptors for the sugar polymers, although the significance of these covalent modifications is unclear at present.

Hyperstimulation of poly(ADP-ribose) polymerase as a consequence of overwhelming DNA damage has been evoked as a mechanism for cellular euthanasia (Gaal et al. 1987). This mechanism involves depletion of NAD$^+$ pools, ultimately leading to an inhibition of oxidative phosphorylation, ATP depletion, and cell death. The response has been proposed as the cause of cell death in certain lymphoid cells exposed to glucocorticoid hormone (Berger et al. 1987) and in cells exposed to hydrogen peroxide (Schraufstatter et al. 1986), a model that may be relevant to macrophage-mediated cell killing.

Most available information on the role of poly(ADP-ribosylation) in thymocyte apoptosis has come from studies using competitive inhibitors, such as nicotinamide and 3-amino-benzamide. Umansky and colleagues have shown that these agents potentiate DNA fragmentation in response to gamma irradiation (Nelipovich et al. 1988), and they concluded that inhibition of poly(ADP-ribose) polymerase may be a general mechanism for endonuclease activation during apoptosis. We have also found evidence for the involvement of the enzyme in DNA fragmentation in rat liver nuclei, although in that system, poly(ADP-ribose) polymerase activity appeared to be a requirement for endonuclease activation (Jones et al. 1989). However, we were unable to link cell death to NAD^+ depletion in glucocorticoid- or Ca^{++} ionophore-treated thymocytes, and 3-aminobenzamide had no protective effect on the cells (McConkey et al. 1989b). The link between this possible "suicide program" and apoptosis therefore requires clarification, and further investigation will be required to determine the molecular basis for the interaction between poly(ADP-ribose) polymerase and the endogenous endonuclease.

The roles of Ca^{++}, PKC, and cAMP/PKA in the regulation of thymocyte apoptosis are summarized in Figure 2. Taken together, these observations suggest that the signaling pathways involved in T-cell proliferation and apoptosis are well conserved. Moreover, that Ca^{++} or cAMP can induce either thymocyte apoptosis or peripheral clonal anergy strongly suggests that the mechanisms involved in the maintenance of tolerance are completely conserved. Developmental status alone may dictate whether these signals result in death or proliferative nonresponsiveness. More recent work showing that peripheral T cells can be deleted (Jones et al. 1990), apparently by apoptosis, suggests that the state of hypersensitivity or insensitivity to apoptosis is reversible. This reversibility may itself be an important physiological control mechanism.

Model for Thymic Cell Selection

Intrathymic cell selection promotes the survival of thymocytes with TCRs that can recognize foreign antigens in the context of self-MHC molecules, while removing those cells bearing auto-reactive antigen receptors. The available information indicates

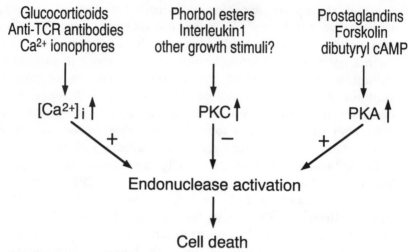

FIGURE 2 Cellular signaling in thymocyte apoptosis.

that the TCR is involved in both positive and negative thymocyte selection. It appears likely that TCR-stimulated apoptosis mediates clonal deletion of autoreactive cells. Other mechanisms of tolerance induction exist, including TCR-, glucocorticoid-, and cAMP-mediated induction of clonal anergy in mature thymocytes and peripheral T cells.

The observations discussed above lend themselves well to an overall hypothesis for thymocyte education, outlined schematically in Figure 3. In stage I, bone-marrow-derived stem cells home to the thymus. Following events that likely involve cell-cell interactions, thymic precursors acquire a number of differentiation markers, including the IL-2 receptor. These progenitors expand by an IL-2-dependent mechanism until a differentiation signal causes IL-2 production to shut down, resulting in loss of IL-2 receptor expression, expression of the TCR, CD4, and CD8. A precedent for induction of sensitivity to apoptosis can be obtained from studies of the effects of growth factor withdrawal on a wide variety of growth-factor-dependent cells, including testosterone-dependent cells of the prostate (Kyprianou and Isaacs 1988; Kyprianou et al. 1988), IL-3-dependent hematopoietic stem cells (Williams et al. 1990), and IL-2-dependent T lymphocytes (Duke and Cohen 1986). There-

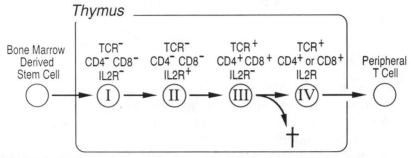

FIGURE 3 Hypothetical model for thymic cell selection. Stages of intrathymic development are (I) commitment; (II) expansion, TCR gene recombination; (III) selection; and (IV) immunocompetence. (IL-2R) Receptor for interleukin 2. For a more detailed discussion of what is known about the significance of surface phenotype in thymic cell selection, see Blackman et al. (1990).

fore, it is certainly conceivable that immature stage II thymocytes become "addicted" to IL-2 prior to acquisition of CD4 and CD8, and that IL-2 withdrawal causes them to become apoptosis-sensitive.

Accessory cells in the thymus then determine the fate of clones within the stage III, CD4+CD8+ thymocyte pool. Apoptosis occurs when CD4+CD8+ thymocytes encounter accessory cells displaying self-antigens due to the lack of a second signal required for activation. In positive selection, this second signal may involve a growth factor (IL-1), binding to a coreceptor (CD4,CD8), or both, and the second signal may serve to augment TCR-mediated activation of PKC. Responsiveness to IL-2, loss of either CD4 or CD8 expression, and desensitization to apoptosis would accompany positive selection of the stage III cells.

In mature stage IV thymocytes, another tolerance mechanism may be induced by accessory cells. Analogous to TCR-mediated apoptosis in stage III cells, unbalanced TCR signaling may lead to anergy in the mature thymocyte pool, thereby achieving tolerance without deletion. Once these cells emigrate to the periphery, tolerance may be maintained by the same type of interaction. Alternatively, sensitivity to apoptosis may be reinduced by IL-2 withdrawal, and deletion of unwanted (or excess in the case of cells that have undergone clonal expansion) T lymphocytes could be induced by TCR-mediated apoptosis.

CONCLUDING REMARKS

In this chapter, we have focused on the signaling processes involved in thymocyte apoptosis, but several of the concepts are relevant to apoptosis in other tissues. Calcium-stimulated endogenous endonuclease is found in nuclei isolated from a variety of tissues (Vanderbilt et al. 1982). A link between cAMP and apoptosis in other cell types has been postulated for years (Basile et al. 1973; Daniel et al. 1973; Pratt and Martin 1975). Cell death induced by cytotoxic T lymphocytes (CTL) and natural killer (NK) cells is another well-characterized example of apoptosis (Russell et al. 1982; Duke et al. 1983, 1986) that is discussed in detail elsewhere (Duke, this volume). Although this rapid apoptosis likely involves processes not discussed above, a sustained Ca^{++} increase occurs in CTL or NK cell targets that precedes and may mediate endonuclease activation (Poenie et al. 1987; McConkey et al. 1990b). A general role for PKC in inhibiting apoptosis is suggested by the observations that phorbol esters and other tumor promoters inhibit apoptosis in irradiated cells (Tomei et al. 1988), the liver (Bursch et al. 1984; Columbano et al. 1984), and human synovial cells exposed to cold shock (Perotti et al. 1990). Future work may reveal other examples of apoptosis where these signaling processes are relevant. In addition, ongoing investigation will no doubt turn up additional physiological mechanisms that control the process.

ACKNOWLEDGMENTS

D.J.M. was supported by a fellowship from the Cancer Research Institute and by a grant from the Swedish Medical Research Council (Project No. 03X-2471).

REFERENCES

Afanasyev, V.N., B.A. Korol', Y.A. Mantsygin, P.A. Nelipovich, V.A. Pechatnikov, and S.R. Umansky. 1986. Flow cytometry and biochemical analysis of DNA degradation characteristic of two types of cell death. *FEBS Lett.* **194:** 347.

Basile, D.V., H.N. Wood, and A.C. Braun. 1973. Programming cells for death under defined experimental conditions: Relevance to the tumor problem. *Proc. Natl. Acad. Sci.* **70:** 3055.

Benjamin, R.C. and D.M. Gill. 1980. Poly(ADP-ribose) synthesis in vitro programmed by damaged DNA. *J. Biol. Chem.* **255:** 10493.

Berger, N.A., S.J. Berger, D.C. Sudar, and C.W. Distelhorst. 1987. Role of nicotinamide adenine dinucleotide and adenosine triphosphate in glucocorticoid-induced cytotoxicity in susceptible lymphoid cells. *J. Clin. Invest.* **79:** 1558.

Berridge, M.J. 1984. Inositol trisphosphate and diacylglycerol as second messengers. *Biochem. J.* **220:** 345.

Berridge, M.J. and R.F. Irvine. 1984. Inositol trisphosphate, a novel second messenger in cellular signal transduction. *Nature* **312:** 315.

Blackman, M., J. Kappler, and P. Marrack. 1990. The role of the T cell receptor in positive and negative selection of developing T cells. *Science* **248:** 1335.

Boyer, P.D. and E.V. Rothenberg. 1989. Changes in inducibility of IL-2 receptor α chain and T cell receptor expression during thymocyte differentiation in the mouse. *J. Immunol.* **142:** 4121.

Bursch, W., B. Lauer, I. Timmermann-Trosiener, G. Barthel, J. Schupler, and R. Schulte-Hermann. 1984. Controlled death (apoptosis) of normal and putative preneoplastic cells in rat liver following withdrawal of tumor promoters. *Carcinogenesis* **5:** 453.

Calder, V., S. Owen, C. Watson, M. Feldmann, and A. Davison. 1989. MS: A localized immune disease of the central nervous system. *Immunol. Today* **10:** 99.

Cohen, J.J. and R.C. Duke. 1984. Glucocorticoid activation of a calcium-dependent endonuclease in thymocyte nuclei leads to cell death. *J. Immunol.* **132:** 38.

Columbano, A., G.M. Ledda-Columbano, P.M. Rao, S. Rajalakshmi, and D.S.R. Sarma. 1984. Occurrence of cell death (apoptosis) in preneoplastic and neoplastic liver cells. A sequential study. *Am. J. Pathol.* **116:** 441.

Crabtree, G.R. 1989. Contingent genetic regulatory events in T lymphocyte activation. *Science* **243:** 355.

Daniel, V., G. Litwack, and G.M. Tomkins. 1973. Induction of cytolysis of cultured lymphoma cells by adenosine 3′:5′-cyclic monophosphate and the isolation of resistant variants. *Proc. Natl. Acad. Sci.* **70:** 76.

Denner, L.A., N.L. Weigel, B.L. Maxwell, W.T. Schrader, and B.W. O'Malley. 1990. Regulation of progesterone receptor-mediated transcription by phosphorylation. *Science* **250:** 1740.

Duke, R.C. and J.J. Cohen. 1986. IL-2 addiction: Withdrawal of growth factor activates a suicide program in dependent T cells. *J.*

Lymphokine Res. **5**: 289.

Duke, R.C., R. Chervenak, and J.J. Cohen. 1983. Endogenous endonuclease-induced DNA fragmentation: An early event in cell-mediated cytolysis. *Proc. Natl. Acad. Sci.* **80**: 6361.

Duke, R.C., J.J. Cohen, and R. Chervenak. 1986. Differences in target cell DNA fragmentation induced by mouse cytotoxic T lymphocytes and natural killer cells. *J. Immunol.* **137**: 1442.

Durum, S.K., J.A. Schmidt, and J.J. Oppenheim. 1985. Interleukin 1: An immunological perspective. *Annu. Rev. Immunol.* **3**: 263.

Fowlkes, B.J. and D.M. Pardoll. 1989. Molecular and cellular events of T cell development. *Adv. Immunol.* **44**: 207.

Gaal, J.C., K.R. Smith, and C.K. Pearson. 1987. Cellular euthanasia mediated by a nuclear enzyme: A central role for nuclear ADP-ribosylation in cellular metabolism. *Trends Biochem. Sci.* **12**: 129.

Goodwin, J.S., D. Atluru, S. Sierakowski, and L.A. Lianos. 1986. Mechanism of action of glucocorticosteroids: Inhibition of T cell proliferation and interleukin 2 production by hydrocortisone is reversed by leukotriene B_4. *J. Clin. Invest.* **77**: 1244.

Gruol, D.J., N.F. Campbell, and S. Bourgeous. 1986. Cyclic AMP-dependent protein kinase promotes glucocorticoid receptor function. *J. Biol. Chem.* **261**: 4909.

Imboden, J.B. and A. Weiss. 1987. The T-cell antigen receptor regulates sustained increases in cytoplasmic free Ca^{2+} through extracellular Ca^{2+} influx and ongoing intracellular Ca^{2+} mobilization. *Biochem. J.* **247**: 695.

Jenkins, M., D.M. Pardoll, J. Mizuguchi, T.M. Chused, and R.H. Schwartz. 1987. Molecular events involved in the induction of a nonresponsive state in interleukin 2-producing helper T lymphocyte clones. *Proc. Natl. Acad. Sci.* **84**: 5409.

Johnson, K.W., B.H. Davis, and K.A. Smith. 1988. cAMP antagonizes interleukin 2-promoted T-cell cycle progression at a discrete point in early G_1. *Proc. Natl. Acad. Sci.* **85**: 6072.

Johnstone, A.P. and G.T. Williams. 1982. Role of DNA breaks and ADP-ribosyl transferase activity in eukaryotic differentiation demonstrated in human lymphocytes. *Nature* **300**: 368.

Jones, L.A., L.T. Chin, D.L. Longo, and A.M. Kruisbeek. 1990. Peripheral clonal elimination of functional T cells. *Science* **250**: 1726.

Jones, D.P., D.J. McConkey, P. Nicotera, and S. Orrenius. 1989. Calcium-activated DNA fragmentation in rat liver nuclei. *J. Biol. Chem.* **264**: 6398.

Kaiser, N. and I.S. Edelman. 1977. Calcium dependence of glucocorticoid-induced lymphocytolysis. *Proc. Natl. Acad. Sci.* **74**: 638.

———. 1978. Further studies on the role of calcium in glucocorticoid-induced lymphocytolysis. *Endocrinology* **103**: 936.

Kammer, G.M. 1988. The adenylate cyclase-cAMP-protein kinase A pathway and regulation of the immune response. *Immunol. Today* **9:** 222.

Kizaki, H., H. Shimada, F. Ohsaka, and T. Sakurada. 1988. Adenosine, deoxyadenosine, and deoxyguanosine induce DNA cleavage in mouse thymocytes. *J. Immunol.* **141:** 1652.

Kristensen, L.W., F. Bettens, and A.L. deWeck. 1983. Lymphokine regulation of activated (G_1) lymphocytes. I. Prostaglandin E_2-induced inhibition of interleukin 2 production. *J. Immunol.* **130:** 1770.

Kyprianou, N. and J.T. Isaacs. 1988. Activation of programmed cell death in the rat ventral prostate after castration. *Endocrinology* **122:** 552.

Kyprianou, N., H.F. English, and J.T. Isaacs. 1988. Activation of a Ca^{2+}-Mg^{2+}-dependent endonuclease as an early event in castration-induced prostatic cell death. *Prostate* **13:** 103.

Leiter, E.H. 1989. The genetics of diabetes susceptibility in mice. *FASEB J.* **3:** 2231.

Lockshin, R.A. 1981. Cell death in metamorphosis. In *Cell death in biology and pathology* (ed. I.D. Bowen and R.A. Lockshin), p. 79. Chapman and Hall, London.

Lunec, J. 1984. Introductory review: Involvement of ADP-ribosylation in cellular recovery from some forms of DNA damage. *Br. J. Cancer* **49:** 13.

McConkey, D.J., P. Hartzell, and S. Orrenius. 1990a. Rapid turnover of endogenous endonuclease activity in thymocytes: Effects of inhibitors of macromolecular synthesis. *Arch. Biochem. Biophys.* **278:** 284.

McConkey, D.J., S. Orrenius, and M. Jondal. 1990b. Cellular signalling in programmed cell death. *Immunol. Today* **11:** 120.

——. 1990c. Agents that elevate cAMP stimulate DNA fragmentation in thymocytes. *J. Immunol.* **145:** 1227.

——. 1990d. NK cell-induced cytotoxicity is dependent on a Ca^{2+} increase in the target. *FASEB J.* **4:** 2661.

McConkey, D.J., P. Hartzell, M. Jondal, and S. Orrenius. 1989a. Inhibition of DNA fragmentation in thymocytes and isolated thymocyte nuclei by agents that stimulate protein kinase C. *J. Biol. Chem.* **264:** 13399.

McConkey, D.J., P. Hartzell, P. Nicotera, and S. Orrenius. 1989b. Calcium-activated DNA fragmentation kills immature thymocytes. *FASEB J.* **3:** 1843.

McConkey, D.J., P. Hartzell, J.F. Amador-Perez, S. Orrenius, and M. Jondal. 1989c. Calcium-dependent killing of immature thymocytes by stimulation via the CD3/T cell receptor complex. *J. Immunol.* **143:** 1801.

McConkey, D.J., P. Hartzell, S.C. Chow, S. Orrenius, and M. Jondal. 1990e. Interleukin 1 inhibits T cell receptor-mediated apoptosis in immature thymocytes. *J. Biol. Chem.* **265:** 3009.

McConkey, D.J., P. Hartzell, S.K. Duddy, H. Hakansson, and S. Orrenius. 1988. 2,3,7,8-Tetrachlorodibenzo-*p*-dioxin kills immature thymocytes by Ca^{2+}-mediated endonuclease activation. *Science* **242:** 256.

McConkey, D.J., P. Nicotera, P. Hartzell, G. Bellomo, A.H. Wyllie, and S. Orrenius. 1989d. Glucocorticoids activate a suicide process in thymocytes through an elevation of cytosolic Ca^{2+} concentration. *Arch. Biochem. Biophys.* **269:** 363.

McDuffie, M., W. Born, P. Marrack, and J. Kappler. 1986. The role of T-cell receptor in thymocyte maturation: Effects of in vivo anti-receptor antibody. *Proc. Natl. Acad. Sci.* **83:** 8728.

Mercep, M., A.M. Weissman, S.J. Frank, R.D. Klausner, and J.D. Ashwell. 1989. Activation-driven cell death and T cell receptor zeta-eta expression. *Science* **246:** 1162.

Mercep, M., J.S. Bonifacino, P. Garcia-Morales, L.E. Samelson, R.D. Klausner, and J.D. Ashwell. 1988. T cell CD3-zeta/eta heterodimer expression and coupling to phosphoinositide hydrolysis. *Science* **242:** 571.

Meuer, S.C., R.E. Hussey, D.A. Cantrell, J.C. Hodgdon, S.F. Schlossman, K.A. Smith, and E.L. Reinherz. 1984. Triggering of the T3-Ti antigen-receptor complex results in clonal T-cell proliferation through an interleukin 2-dependent autocrine pathway. *Proc. Natl. Acad. Sci.* **81:** 1509.

Murphy, K.M., A.B. Heimberger, and D.Y. Loh. 1990. Induction by antigen of intrathymic apoptosis of $CD4^+CD8^+$ TCR^{lo} thymocytes in vivo. *Science* **250:** 1720.

Mustelin, T., K.M. Coggeshall, N. Isakov, and A. Altman. 1990. T cell receptor-mediated activation of phospholipase C requires tyrosine phosphorylation. *Science* **247:** 1584.

Nelipovich, P.A., L.V. Nikonova, and S.R. Umansky. 1988. Inhibition of poly(ADP-ribose) polymerase as a possible reason for activation of Ca^{2+}/Mg^{2+} endonuclease in thymocytes of irradiated rats. *Int. J. Radiat. Biol.* **53:** 749.

Nishizuka, Y. 1984. The role of protein kinase C in cell surface signal transduction and tumour promotion. *Nature* **308:** 693.

Perotti, M., F. Toddei, F. Mirabelli, M. Vairetti, G. Bellomo, D.J. McConkey, and S. Orrenius. 1990. Calcium-dependent DNA fragmentation in human synovial cells exposed to cold shock. *FEBS Lett.* **259:** 331.

Poenie, M., R.Y. Tsien, and S.-M. Schmitt-Verhulst. 1987. Sequential activation and lethal hit measured by $[Ca^{2+}]_i$ in individual cytolytic T cells and targets. *EMBO J.* **6:** 2223.

Pratt, R.M. and G.R. Martin. 1975. Epithelial cell death and cyclic AMP increase during palatal development. *Proc. Natl. Acad. Sci.* **72:** 874.

Rammensee, H.-G., R. Kroschewski, and B. Frangoulis. 1989. Clonal anergy induced in mature $VB6^+$ T lymphocytes on immunizing

Mls-1^b mice with Mls-1^a expressing cells. *Nature* **339:** 541.

Rosoff, P.M., N. Savage, and C.A. Dinarello. 1988. Interleukin 1 stimulates diacylglycerol production in T lymphocytes by a novel mechanism. *Cell* **54:** 73.

Ross, C.A., J. Meldolesi, T.A. Milner, T. Satoh, S. Supattapone, and S.H. Snyder. 1989. Inositol 1,4,5-trisphosphate receptor localized to endoplasmic reticulum in cerebellar Purkinje neurons. *Nature* **339:** 468.

Russell, J.H., V. Masakowski, T. Rucinsky, and G. Phillips. 1982. Mechanisms of immune lysis. III. Characteristics of the nature and kinetics of the cytotoxic T lymphocyte-induced nuclear lesion in the target. *J. Immunol.* **128:** 2087.

Schraufstatter, I.U., D.B. Hinshaw, P.A. Hyslop, R.G. Spragg, and C.G. Cochrane. 1986. Oxidant injury of cells: DNA strand breaks activate polyadenosine diphosphate-ribose polymerase and lead to depletion of nicotinamide adenine dinucleotide. *J. Clin. Invest.* **77:** 1312.

Schwartz, R.H. 1990. A cell culture model for T lymphocyte clonal anergy. *Science* **248:** 1349.

Sellins, K.S. and J.J. Cohen. 1987. Gene induction by gamma irradiation leads to DNA fragmentation in lymphocytes. *J. Immunol.* **139:** 3199.

Shi, Y., B.M. Sahai, and D.R. Green. 1989. Cyclosporin A inhibits activation-induced cell death in T-cell hybridomas and thymocytes. *Nature* **339:** 625.

Shi, Y., M.G. Szalay, L. Paskar, M. Boyer, B. Singh, and D.R. Green. 1990. Activation-induced cell death in T cell hybridomas is due to apoptosis. Morphologic aspects and DNA fragmentation. *J. Immunol.* **144:** 3326.

Smith, C.A., G.T. Williams, R. Kingston, E.J. Jenkinson, and J.J.T. Owen. 1989. Antibodies to CD3/T-cell receptor complex induce death by apoptosis in immature T cells in thymic cultures. *Nature* **337:** 181.

Tomei, L.D., P. Kanter, and C.E. Wenner. 1988. Inhibition of radiation-induced apoptosis in vitro by tumor promoters. *Biochem. Biophys. Res. Comm.* **155:** 324.

Tonegawa, S. 1983. Somatic generation of antibody diversity. *Nature* **302:** 575.

Truneh, A., F. Albert, P. Golstein, and A.-M. Schmitt-Verhulst. 1985. Early steps of lymphocyte activation bypassed by synergy between calcium ionophores and phorbol esters. *Nature* **313:** 318.

Umansky, S.R., B.A. Korol, and P.A. Nelipovich. 1981. In vivo DNA degradation in thymocytes of gamma-irradiated or hydrocortisone-treated rats. *Biochem. Biophys. Acta* **655:** 9.

Vanderbilt, J.N., K.S. Bloom, and J.N. Anderson. 1982. Endogenous nuclease: Properties and effects on transcribed genes in chromatin. *J. Biol. Chem.* **257:** 13009.

White, J., A. Herman, A.-M. Pullen, R. Kubo, J.W. Kappler, and P. Marrack. 1989. The VB-specific superantigen staphylococcal enterotoxin B: Stimulation of mature T cells and clonal deletion in neonatal mice. *Cell* **56**: 27.

Williams, G.T., C.A. Smith, E. Spooncer, T.M. Dexter, and D.R. Taylor. 1990. Haemopoietic colony stimulating factors promote cell survival by suppressing apoptosis. *Nature* **343**: 76.

Wyllie, A.H. 1980. Glucocorticoid-induced thymocyte apoptosis is associated with endogenous endonuclease activation. *Nature* **284**: 555.

Wyllie, A.H., J.F.R. Kerr, and A.R. Currie. 1980. Cell death: The significance of apoptosis. *Int. Rev. Cytol.* **68**: 251.

Wyllie, A.H., R.G. Morris, A.L. Smith, and D. Dunlop. 1984. Chromatin cleavage in apoptosis: Association with condensed chromatin morphology and dependence on macromolecular synthesis. *J. Pathol.* **142**: 67.

Zuniga-Pflucker, J.C., L.A. Jones, D.C. Longo, and A.M. Kruisbeek. 1990. CD8 is required during positive selection of CD4-/CD8+ T cells. *J. Exp. Med.* **171**: 427.

Zuniga-Pflucker, J.C., S.A. McCarthy, M. Weston, D.L. Longo, A. Singer, and A.M. Kruisbeek. 1989. Role of CD4 in thymocyte maturation and selection. *J. Exp. Med.* **169**: 2085.

Programmed Cell Death in the Peripheral Nervous System

D.P. Martin and E.M. Johnson, Jr.
Department of Molecular Biology and Pharmacology
Washington University School of Medicine
St. Louis, Missouri 63110

Neurogenesis produces about twice as many neurons as are needed by the mature nervous system (Cowan et al. 1984). This initial overabundance is subsequently pruned by the natural death of about half of these neurons, providing a mechanism whereby the size of the neuronal pool is matched to the amount of target tissue to be innervated. This interpretation is supported by the fact that neuronal death is reduced by increasing the size of the target and is exacerbated by removing the target.

Neuronal death often follows target removal, not because of the initial trauma of axotomy, but rather because the neuron becomes deprived of target-derived influences that are necessary for survival. One such target-derived influence consists of a class of proteins known as neurotrophic factors, the best-characterized member of which is nerve growth factor (NGF). NGF is produced by the targets of sympathetic and sensory neurons derived from the neural crest, as well as by the targets of cholinergic neurons of the basal forebrain (Thoenen et al. 1987). Neurons die if they receive inadequate neurotrophic support. For example, dependent neurons die when endogenous NGF is neutralized by injecting antibodies to the factor (Levi-Montalcini and Booker 1960). Conversely, neuronal death can be prevented after axotomy if exogenous NGF is provided (Hendry and Campbell 1976; Hefti 1988).

Over the past 30 years, it has become apparent that NGF, when added to responsive cells, affects many cellular events. An implicit assumption in these investigations has been that NGF "supports" life in some fashion by stimulating anabolic

processes within the neuron. Although NGF does indeed produce "trophic" or "nourishing" effects consistent with this
hypothesis, no result has emerged that explains why NGF is
required for neuronal survival. We chose to address this question from a different viewpoint; rather than asking what NGF
does when added to a cell, we asked why neurons die when
NGF is removed. Herein, we review data indicating that NGF
supports neuronal survival by suppressing an active suicide
program rather than by trophically stimulating anabolic processes.

Morphology of Neuronal Death

Our studies were conducted with an in vitro model of neuronal
death in which cultures of sympathetic rat neurons were established in the presence of NGF for 1 week and then acutely
deprived of trophic support by adding anti-NGF serum (Martin
et al. 1988). This experimental paradigm mimics the physiological situation encountered by neurons during development
or after axotomy, when trophic support becomes insufficient
and neurons die.

When NGF was removed from 7-day-old cultures by adding
anti-NGF to the culture medium, no changes were apparent
for the first 18–24 hours. Then the neurites became thinner
and disrupted in places, leaving behind bits of neuritic debris.
By 30 hours, some cell bodies were smaller and phase dark.
There were distinctly fewer recognizable neurons on the dish
at 36 hours, although no debris was seen floating in the culture medium. The intact cells that were observed at this time
were frequently alone, surrounded by scattered, phase-dark
debris. By 48 hours, only a few phase-bright neurons could be
found, isolated amid the remnants of cell bodies and neuritic
debris. This debris remained attached to the culture dish for
several days before lifting up and floating into the medium.

Time-lapse phase-contrast video observation of dying sympathetic neurons indicated that the death of each neuron was
a rapid process, transpiring over tens of minutes. As the
neurons died, dramatic cytoplasmic oscillations were sometimes observed, characteristic of the process described as
zeiosis. Our impression was that death commenced in each

neuron at slightly different times between 24 and 48 hours postdeprivation, making a temporal description of the ultrastructural events difficult.

Characterization of neuronal death with electron microscopy revealed that untreated neurons exhibited large vesicular nuclei with finely dispersed chromatin, large prominent nucleoli, cytoplasm filled with polysomes and rough endoplasmic reticulum, a complement of thread-like mitochondria, and, occasionally, one or two lipid droplets. The interneuronal neuropil was composed of large numbers of small, normal-appearing neurites with a prominent cytoskeleton.

The first detectable ultrastructural alteration in cultures subjected to NGF deprivation occurred 12–18 hours after the addition of anti-NGF and consisted of increased numbers of abnormal or actively degenerating neurites. The ultrastructural appearance of the axonopathy was represented by a continuum of degenerative alterations, ranging from swollen axons containing numerous intra-axonal multivesicular bodies or dense degenerating organelles to frank loss of axonal integrity resulting in the release of axonal debris into the culture medium. Although many degenerating axons were adjacent to neurons, degeneration did not involve axonal termini exclusively. There was no clear evidence of structural alterations involving the neuronal perikarya at the earliest time of distinct axonopathy (roughly 12–18 hours); however, subtle changes in the perikarya began to develop 18–24 hours after NGF deprivation. The earliest neuronal cell body alterations consisted of (1) slight shrinkage of the nucleus, which resulted in an irregular nuclear perimeter, diffuse increase in the density of chromatin, and development of small patches of heterochromatin and (2) appearance of small lipid droplets, occurring singly or in small groups in the perikaryal cytoplasm. At this time, polysomes and rough endoplasmic reticulum, mitochondria, and neuronal cytoskeletal elements were well preserved.

Most neuronal perikarya began to develop significant ultrastructural alterations between 24 and 30 hours. The degree of axonal degeneration increased markedly with prominent axonal dissolution. Few normal neurites remained. Neurons developed marked nuclear irregularity and apparent shrinkage without the development of large discrete nuclear heterochromatin aggregates. Surviving neurons gave rise to a few neuritic processes that may represent the residue of preferential pruning of axonal processes at earlier times. Increased numbers of lipid inclusions were found in the neuronal cell body, which occasionally resulted in the marked degree of lipid accumulation seen in a few neurons. Accumulated lipid droplets typically lacked substructure and a limiting membrane, although marked accumulation of lipid droplets displaced normal cytoplasmic organelles, especially rough endoplasmic reticulum, the total amount of which appeared diminished in comparison with controls. In addition, the cisternae of the rough endoplasmic reticulum were often dilated by a dense granular material. The majority of ribosomes remained attached to rough endoplasmic reticulum or represented cytoplasmic polysomes. A few bundles of neurofilaments were also encountered in the perikaryal cytoplasm.

The final phases of neuronal degeneration began substantively at 30 hours and were advanced at later times. Several patterns of degeneration were observed. The first and most common degenerative pattern was the formation of swollen clear vacuoles admixed with degenerating organelles in lucent perikaryal cytoplasm, which eventually resulted in the dissolution of the nuclear and plasma membranes and the liberation of intracellular contents into the culture medium. Another degenerative pattern, although infrequent, was characterized by the condensation of nucleoplasm and cytoplasm forming dense cytoplasmic protrusions of the neuronal plasmalemma containing fragments of cytoplasmic and nuclear debris. It was difficult to assign a precise longitudinal sequence of morphological events associated with neuronal death because, at later times, all stages of death were observed. Neuronal perikarya showed no morphological evidence of autophagy.

The significance of lipid droplet formation by neurons during NGF deprivation remains uncertain. The inclusion of lipid droplets in the cell body reflects altered lipid metabolism or transport. One possibility is that the neuritic membrane accumulates at a rate exceeding degradative capability with the excess amassing as droplets in the neuronal perikarya. The observed depression of protein synthesis at the times of this lipid buildup may result in insufficient amounts of lipoproteins required for lipid export. Alternatively, the droplets may reflect lipid that is newly synthesized in excess of the rate at which it can be used. The latter possibility seems less likely because we observed lipid droplets to accumulate progressively as cells die. Abundant lipid droplet formation in NGF-deprived sympathetic neurons has not been observed in vivo (Levi-Montalcini et al. 1969; Angeletti et al. 1971; Schucker 1972; Pannese 1976), suggesting that it is not pathognomonic of NGF deprivation. Perhaps the in vivo environment provides interactions that limit neuronal lipid loading, such as hematogenously derived phagocytic cells that may engulf excess neuronal lipid and other debris.

The fact that nuclei and nucleoli tended to remain normal until late stages of death suggests that they might be participating in ongoing RNA and protein synthesis. A minority of degenerating neurons showed an ultrastructural appearance consistent with the form of death known as "apoptosis" (Wyllie et al. 1980), which has been suggested as a common morphological motif of natural cell death, although morphological evaluation of programmed cell deaths from various other systems has failed to indicate a common biochemical mechanism (Clarke 1990). The accumulation of neutral lipid is often associated with necrosis, rather than apoptosis. Curiously, however, somatic cells in the alga, *Volvox carteri*, accumulate lipid droplets as they senesce and die (Pommerville and Kochert 1981). Another morphological distinction is the apparent absence of massive chromatin condensation and margination in NGF-deprived neurons.

Neuronal Death Is an Active Process

If neurons die because they are "malnourished" and suffer a passive demise from lack of trophic stimulation, the prediction would be that the neurons would die faster than normal after inhibition of macromolecular synthesis. Conversely, if cell death is an active process resulting from the products of a "death program," it would be predicted that inhibiting macromolecular synthesis should prevent cell death. Inhibiting RNA and protein synthesis *prevents* the neurons from dying (Martin et al. 1988). This experiment is repeated as a control in Figure

1, which shows that cycloheximide (CHX) and actinomycin D (AMD) prevent the release of the intracellular enzyme adenylate kinase (AK). AK is released as cells die and thus indicates the neuronal destruction caused by NGF deprivation.

Since neuronal death was prevented by these inhibitors rather than accelerated, the death of NGF-deprived neurons must result from the activity of a new set of proteins or the altered (e.g., increased) activity of existing proteins that are capable of killing the cell in the stereotypic manner described here. We shall refer to this new or augmented set of activities collectively as the death program, involving a cascade of new mRNA and protein synthesis, which results in the production of "killer protein(s)." For the proximate killer protein(s), we suggest the term "thanatins" (after Thanatos, the Greek god of

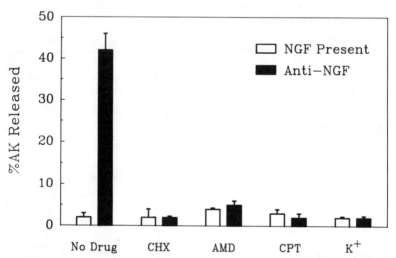

FIGURE 1 Inhibitors of protein or RNA synthesis, as well as CPT-cAMP and depolarization, prevent neuronal death caused by NGF deprivation. Sympathetic neurons were grown in the presence of NGF for 1 week and then exposed to the following drugs in the presence (open bar) or absence (filled bar) of NGF. (No Drug) Normal culture medium; (CHX) 1 µg/ml cycloheximide; (AMD) 0.1 µg/ml actinomycin D; (CPT) 300 µM chlorophenylthio-cAMP; (K+) 35 mM potassium chloride. After 60 hr, the amount of AK released into the medium by dying cells was measured and expressed as a percentage relative to the total activity present in each well. Each bar represents the mean ± S.D. of three wells.

death). Thus, inhibitors of protein and mRNA synthesis non-selectively mimic the ability of NGF to suppress the expression of these genes and gene products.

Oppenheim et al. (1990) have demonstrated that inhibitors of RNA and protein synthesis prevent the naturally occurring death of motor and dorsal root ganglion sensory neurons in the chicken embryo. Naturally occurring neuronal death also requires protein synthesis in the insect nervous systems of *Manduca* (Fahrbach and Truman 1988) and *Drosophila* (Kimura and Truman 1990). Our initial observation has been extended to other neuronal types in vitro. Scott and Davies (1990) have demonstrated that RNA and protein synthesis inhibitors prevent neuronal death after trophic factor deprivation of (1) chicken trigeminal mesencephalic neurons, a population of proprioceptive neurons supported by brain-derived neurotrophic factor (BDNF) (Davies et al. 1986) but not NGF (Davies et al. 1987); (2) dorsomedial trigeminal ganglion neurons, a population of cutaneous sensory neurons supported by NGF; and (3) ciliary ganglion neurons, a population of parasympathetic neurons supported by ciliary neurotrophic factor (CNTF) (Barbin et al. 1984), but not NGF or BDNF (Lindsay et al. 1985). Thus, observations consistent with a death program have been described for at least four different neuronal populations supported by at least three different neurotrophic factors. Therefore, we suggest the generalization that suppression of the death program is the mechanism by which all neurotrophic factors promote survival of their respective neuronal populations.

An active death program may be expressed in other cell types when conditions are appropriate for their physiological death. For example, protein and RNA synthesis are required for the death of prostate epithelium caused by castration (Kyprianou and Isaacs 1988), and the death of intersegmental muscles of the saturniid silk moth is blocked by inhibition of protein and RNA synthesis (Lockshin 1969). Additionally, in the nematode, *Caenorhabditis elegans*, two genes have been identified that are specifically required for naturally occurring cell death (Ellis and Horvitz 1986). Recently, these cell death genes have been determined to act autonomously within the dying cells (Yuan and Horvitz 1990).

cAMP

We have confirmed and extended the observations of Rydel and Green (1988), which demonstrate that p-chlorophenylthio cAMP (CPT-cAMP) can maintain the survival of sympathetic neurons in the absence of NGF. Week-old sympathetic neurons do not die when deprived of NGF if CPT-cAMP is present (Fig. 1). Drugs that raise intracellular levels of cAMP, such as cholera toxin, isobutylmethylxanthine, or forskolin, also prevent the death of NGF-deprived neurons. Furthermore, isobutylmethylxanthine, at a dose insufficient in itself to prevent death (100 μM), markedly potentiates the dose response curves of both CPT-cAMP and forskolin, lowering the EC_{50} of each tenfold. Analysis of the intracellular concentration of cAMP fails to reveal any effect of NGF on cAMP concentrations or any cAMP change in association with the death of NGF-deprived neurons. We conclude from this observation that although raised levels of cAMP can pharmacologically repress the death program, cAMP is probably not involved in its physiological regulation by NGF.

Calcium

Depolarization of the membrane potential also prevented neuronal death caused by NGF deprivation. When deprived of NGF in the presence of 35 mM KCl, neurons did not die (Fig. 1). The cholinergic agonist carbachol, as well as high concentrations of choline (acting as a weak cholinergic agonist), also prevented neuronal death. The effect of depolarization was blocked either by withdrawal of extracellular calcium or by chelating intracellular calcium, suggesting that elevated intracellular calcium concentrations were responsible for the saving effect. This also indicates that the neuronal death program does not require elevated calcium concentrations, although calcium does modulate its expression. Experiments with the calcium channel antagonists nimodipine and nifedipine indicated that the influx of extracellular calcium that prevents death under depolarizing conditions is through L-type calcium channels (Koike et al. 1989).

On the basis of these results and studies of the role of calcium as a mediator of excitotoxicity, we propose that for the purpose of neuronal survival, three operational levels of internal free calcium (or calcium in some more subtle pool) can be described. There exists an optimal level of internal calcium that allows for an autonomous existence of the neuron independent of external "trophic" factors. At reduced intracellular calcium concentrations, neurons become dependent on an external trophic factor (in the case of sympathetic neurons, NGF) to suppress the active mechanism that can destroy the neurons (Martin et al. 1988). Excessive influx of calcium, as after excitatory amino acids, may raise intracellular calcium to very high levels, resulting in toxic or fatal effects (Choi 1988).

Treatments That Do Not Prevent Neuronal Death

Inhibitors of RNA and protein synthesis, cAMP, and depolarization are relatively unique in preventing neuronal death. A partial list of treatments that do not alter the death of NGF-deprived neurons is presented in Table 1. For example, it is possible that the destruction of dying neurons subsequent to NGF deprivation is mediated by lysosomal autophagy. We attempted to prevent neuronal death subsequent to NGF deprivation with various inhibitors of lysosomal proteases that included leupeptin, chloroquine, and phenylmethanesulfonyl fluoride (PMSF). None of these agents prevented neuronal death subsequent to NGF deprivation. These observations have tentatively excluded several cellular metabolic pathways from a critical role in neuronal death (e.g., lysosomal proteolysis). This suggests that the death program consists of a relatively well-circumscribed set of activities, rather than global alterations of cellular metabolism.

Model of Neuronal Responses to Trophic Factor Deprivation

Figure 2 illustrates our working hypothesis in schematic form. This model is largely based on the biology of NGF, but the principles outlined should apply generally to other trophic factors. The response to trophic factor begins with its binding to specific receptors located in the neuronal membrane. The trophic factor and receptor are both internalized and retrogradely transported to the cell body. The second messenger

TABLE 1 *AGENTS THAT DO NOT PREVENT NEURONAL DEATH*

Drug	Dose range	Drug action
$2'-5'A_3$	0.02–200 µg/ml	RNase-L activator
3-Aminobenzimide	0.1–10 mM	ADP-ribosylation inhibitor
Adenosine	0.1–1 mM	adenosine receptor agonist
Al^{+++}	0.05–5 mM	—
Allopurinol	0.04–3 mM	xanthine oxidase inhibitor
Amiloride	0.01–0.1 mM	Na^+/H^+ exchange inhibitor
Aphidicolin	10 µM	DNA polymerase inhibitor
Chloroquine	1 µM–10 mM	lysosome stabilizer
Colcemid	25–500 ng/ml	disrupts microtubules
Dexamethasone	0.1–1 µM	glucocorticoid
Difluoromethyl ornithine	0.3–6 mM	ODC inhibitor
Dehydroepiandrosterone	0.1–100 µM	steroid
EGTA	0–6 mM	Ca^{++} chelator
FGF	50–500 ng/ml	growth factor
Ganglioside GM_1	50–450 µg/ml	—
Gensinoside Rb_1	0.03–3 mM	—
IL-1	10^3-10^5 U/ml	lymphokine
IL-2	10–100 U/ml	lymphokine
IL-6	50–250 U/ml	lymphokine
Insulin	0.8–200 µg/ml	—
Janus Green B	0.001–10 µg/ml	mitochondrial poison
K252a	0.25–2.5 µM	kinase inhibitor
La^{+++}	0.4–0.8 mM	—
Leupeptin	1 µM–10 mM	protease inhibitor
Li^+	0.01–10 mM	—
Mepicrine	1–1000 µM	phospholipase-C inhibitor
8-Methoxypsauralin	0.1–1000 µM	—
Nocodazole	0.01–10 mM	disrupts microtubules
Novobiocin	1–1000 µM	—
PMA	30–300 nM	PKC activator
PMSF	1 µM–10 mM	protease inhibitor
Poly I-C	5–50 µg/ml	inducer of IFN
Polymyxin B	10–100 µM	PKC inhibitor
Putrescine	1 µM–10 mM	polyamine
Spermidine	1 µM–10 mM	polyamine
Spermine	1 µM–10 mM	polyamine
TNF	100–1000 U/ml	—
Trifluoroperizine	1–100 µM	Ca^{++}/calmodulin inhibitor
Tunicamycin	0.04–3 µg/ml	glycosylation inhibitor
Zn^{++}	0.4–0.8 mM	—

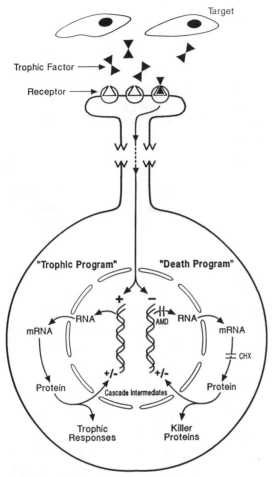

FIGURE 2 Model of trophic factor actions in neurons: Trophic and survival-promoting effects. Neurotrophic factor is released from target tissue, whereupon it binds to specific neuronal receptors and is retrogradely transported to the cell body. The presence of trophic factor stimulates the synthesis of proteins necessary for neurotransmitter production and neuritic outgrowth, i.e., the "trophic program." Trophic factor deprivation, or blockade of its signal transduction by cytosine arabinoside, unleashes a cascade of new mRNA and protein synthesis that ultimately results in killer protein(s) that destroy the neuron, i.e., the "death program." The death program is executed when competition for target during development deprives neurons of trophic factor or when nerve injury separates neurons from their targets. Different cell types may govern the same death program via other physiological regulators.

that carries the NGF signal to the nucleus is unknown. Nonetheless, the information reflecting the presence of NGF becomes available within the nucleus, where it regulates gene expression. We propose that the messenger of trophic factor signal regulates two genetic programs in a complementary fashion. When trophic factor is present, it stimulates the trophic program and represses the death program.

The trophic program consists of a cascade of RNA and protein synthesis that results in the classic trophic responses to NGF, including the induction of neurotransmitter synthesizing enzymes and proteins involved in the production and elongation of neurites. Evidence of a cascade comes from studies on PC12 cells, which show that NGF stimulates an early set of genes, e.g., c-*fos*, followed by the synthesis of other proteins that make up the phenotype of the trophic-supported state.

We hypothesize that the death program is normally repressed by NGF, such that upon NGF deprivation it becomes induced. The death program may consist of a cascade of RNA and protein synthesis, analogous to events such as differentiation and mitosis (Buttyan et al. 1988), resulting in new mRNA that codes for proteins that proximately kill the cell. Although CHX or AMD would be expected to block the trophic and death programs simultaneously, in practice we observed that blockade of protein synthesis in the presence of NGF did not kill neurons for many days. Thus, inhibition of the trophic program is not lethal in the short term. In the absence of NGF, however, these neurons die rapidly. This predicts that inhibiting RNA or protein synthesis should block the death caused by NGF deprivation which is, in fact, what is observed.

CONCLUSIONS

Although neurotrophic factors like NGF may actually nourish responsive cells, loss of trophic *support* is not the reason for neuronal death following trophic factor deprivation. Rather, neurotrophic factor deprivation unleashes a cascade of new mRNA and protein synthesis that ultimately results in killer protein(s) that destroy the neuron. The early components of this cascade may be shared with other cellular events, such as mitosis or differentiation. The death program would be executed when competition for target during development deprives neurons of trophic factor, or after axonal injury that separates the neuron from targets and likewise produces trophic factor deprivation.

Elevated intracellular cAMP and calcium concentrations both prevent the full expression of the death program. Although the significance of these observations is still unclear, the possibility exists that pharmacological manipulation of the death program may allow a rational treatment regimen for nerve trauma or degenerative diseases.

When deprived of trophic support, neurons synthesize new mRNA that ultimately codes for proteins that kill the cell. Different cell types may govern the same death program via other physiological regulators. Thus, an understanding of the death program in neurons may apply to a variety of other situations where cells die either under normal circumstances or pathologically. Identification of components of the death program should allow this hypothesis to be tested.

ACKNOWLEDGMENTS

We thank our many colleagues who contributed the various studies mentioned in this paper. Work from the authors' laboratories was supported by National Institutes of Health grants NS-24679, AG-05681, and GM-07805.

REFERENCES

Angeletti, P.U., R. Levi-Montalcini, and F. Caramia. 1971. Analysis of the effects of the antiserum to the nerve growth factor in adult mice. *Brain Res.* **27**: 343.

Barbin, G., M. Manthorpe, and S. Varon. 1984. Purification of the chick eye ciliary neurotrophic factor. *J. Neurochem.* **43**: 1468.

Buttyan, R., Z. Zakeri, R. Lockshin, and D. Wolgemuth. 1988. Cascade induction of c-*fos*, c-*myc*, and heat shock 70K transcripts during regression of the rat ventral prostate gland. *Mol. Endocrinol.* **2**: 650.

Choi, D.W. 1988. Glutamate neurotoxicity and diseases of the nervous system. *Neuron* **1**: 623.

Clarke, P.G.H. 1990. Developmental cell death: Morphological diversity and multiple mechanisms. *Anat. Embryol.* **181:** 195.

Cowan, W.M., J.W. Fawcett, D.D.M. O'Leary, and B.B. Stanfield. 1984. Regressive events in neurogenesis. *Science* **225:** 1258.

Davies, A.M., A.G.S. Lumsden, and H. Rohrer. 1987. Neural crest-derived proprioceptive neurons express nerve growth factor receptors but are not supported by nerve growth factor in culture. *Neuroscience* **20:** 37.

Davies, A.M., H. Thoenen, and Y.-A. Barde. 1986. Different factors from the central nervous system and periphery regulate the survival of sensory neurons. *Nature* **319:** 497.

Ellis, H.M. and H.R. Horvitz. 1986. Genetic control of programmed cell death in the nematode *C. elegans*. *Cell* **44:** 817.

Fahrbach, S.E. and J.W. Truman. 1988. Cycloheximide inhibits ecdysteroid-regulated neuronal death in the moth *Manduca sexta*. *Soc. Neurosci. Abst.* **14:** 368.

Hefti, F. 1988. Rescue of lesioned septal cholinergic neurons by nerve growth factor: Specificity and requirement for chronic treatment. *J. Neurosci.* **8:** 2986.

Hendry, I.A. and J. Campbell. 1976. Morphometric analysis of rat superior cervical ganglion after axotomy and nerve growth factor treatment. *J. Neurocytol.* **5:** 351.

Kimura, K. and J.W. Truman. 1990. Postmetamorphic cell death in the nervous and muscular systems of *Drosophila melanogaster*. *J. Neurosci.* **10:** 403.

Koike, T., D.P. Martin, and E.M. Johnson, Jr. 1989. Role of Ca^{2+} channels in the ability of membrane depolarization to prevent neuronal death induced by trophic-factor deprivation: Evidence that levels of internal Ca^{2+} determine nerve growth factor dependence of sympathetic ganglion cells. *Proc. Natl. Acad. Sci.* **86:** 6421.

Kyprianou, N. and J.T. Isaacs. 1988. Activation of programmed cell death in the rat ventral prostate after castration. *Endocrinology* **122:** 552.

Levi-Montalcini, R. and B. Booker. 1960. Destruction of the sympathetic ganglia in mammals by an antiserum to the nerve-growth promoting factor. *Proc. Natl. Acad. Sci.* **42:** 384.

Levi-Montalcini, R., F. Caramia, and P.U. Angeletti. 1969. Alterations in the fine structure of nucleoli in sympathetic neurons following NGF-antiserum treatment. *Brain Res.* **12:** 54.

Lindsay, R.M., H. Thoenen, and Y.-A. Barde. 1985. Placode and neural crest-derived sensory neurons are responsive at early developmental stages to brain-derived neurotrophic factor (BDNF). *Dev. Biol.* **112:** 319.

Lockshin, R.A. 1969. Programmed cell death. Activation of lysis by a mechanism involving the synthesis of protein. *J. Insect Physiol.* **15:** 1505.

Martin, D.P., R.E. Schmidt, P.S. DiStefano, O.H. Lowry, J.G. Carter, and E.M. Johnson, Jr. 1988. Inhibitors of protein synthesis and RNA synthesis prevent neuronal death caused by nerve growth factor deprivation. *J. Cell Biol.* **106:** 829.

Oppenheim, R.W., D. Prevette, M. Tytell, and S. Homma. 1990. Naturally occurring and induced neuronal death in the chick embryo *in vivo* requires protein and RNA synthesis: Evidence for the role of cell death genes. *Dev. Biol.* **138:** 104.

Pannese, E. 1976. An electron microscopic study of cell degeneration in chick embryo spinal ganglia. *Neuropathol. Appl. Neurobiol.* **2:** 247.

Pommerville, J.C. and G.D. Kochert. 1981. Changes in somatic cell structure during senescence of *Volvox carteri*. *Eur. J. Cell Biol.* **24:** 236.

Rydel, R.E. and L.A. Greene. 1988. cAMP analogs promote survival and neurite outgrowth in cultures of rat sympathetic and sensory neurons independently of nerve growth factor. *Proc. Natl. Acad. Sci.* **85:** 1257.

Schucker, F. 1972. Effects of NGF-antiserum in sympathetic neurons during early postnatal development. *Exp. Neurol.* **36:** 59.

Scott, S.A. and A.M. Davies. 1990. Inhibition of protein synthesis prevents cell death in sensory and parasympathetic neurons deprived of neurotrophic factor *in vitro*. *J. Neurobiol.* **21:** 630.

Thoenen, H., C. Bandtlow, and R. Heumann. 1987. The physiological function of nerve growth factor in the central nervous system: Comparison with the periphery. *Rev. Physiol. Biochem. Pharmacol.* **109:** 145.

Wyllie, A.H., J.F.R. Kerr, and A.R. Currie. 1980. Cell death: The significance of apoptosis. *Int. Rev. Cytol.* **68:** 251.

Yuan, J. and H.R. Horvitz. 1990. The *Caenorhabditis elegans* genes *ced-3* and *ced-4* act cell autonomously to cause programmed cell death. *Dev. Biol.* **138:** 33.

Neuronal Cell Death and the Role of Apoptosis

A.C. Server[1] and W.C. Mobley[2]
[1]Department of Neurology, Massachusetts General Hospital
Boston, Massachusetts 02114
[2]Department of Neurology, Pediatrics and the Neuroscience Program
University of California, San Francisco, California 94143

Cell death in the nervous system has long been recognized (Collin 1906). The detailed studies of Hamburger and Levi-Montalcini (1949), demonstrating extensive neuron loss in spinal ganglia of the chick embryo, highlighted the significance of cell death for the normal development of the vertebrate nervous system. Evidence of cellular degeneration was detected at the specific developmental stages in which the processes of spinal ganglion neurons contacted their peripheral targets and was far less prevalent in the ganglia that innervated limbs (i.e., the brachial and lumbosacral ganglia) than in those innervating the cervical and thoracic regions. Remarkably, early extirpation of the wing or limb bud produced increased evidence for cellular degeneration in the ganglia innervating the operated region; however, degeneration occurred during the same time frame as during normal development, and the topography of degenerative profiles was the same. It was concluded that differential degeneration of neurons contributed significantly to the histogenesis of the spinal ganglia; the target of neurons was identified as playing a key role in this process (Hamburger and Levi-Montalcini 1949). A number of subsequent reviews (Glücksmann 1951; Saunders 1966) citing evidence of cell loss in the development of different tissues in a variety of organisms emphasized the prominent role of cell death throughout ontogeny. Glücksmann (1951) classified each of the documented cases of cell death in development on the basis of its biological significance. Morphogenetic degenerations occurred

Apoptosis: The Molecular Basis of Cell Death
Copyright 1991 Cold Spring Harbor Laboratory Press 0-87969-366-5/91 $3.00 + 00   **263**

during changes in the forms of organs or tissues; histogenetic degenerations occurred with the differentiation of tissues; and phylogenetic degenerations involved the involution of vestigial or larval organs. Saunders (1966) emphasized the "utility" of regressive phenomena in the normal development of multicellular organisms and placed cell death during embryogenesis within the "same conceptual framework as growth and differentiation."

Although they were remarkably insightful, these early studies failed to stimulate widespread interest in cell death as an important biological phenomenon with a role that extends beyond the period of development into normal adult life, aging, and disease. Moreover, in the absence of an adequate body of experimental data, as well as the appropriate theoretical framework and technology to generate it, these studies were unable to provide significant insight into the mechanisms by which cells die. It is in this context that the work of Kerr, Wyllie, and their collaborators (Kerr et al. 1972; Wyllie et al. 1980; Walker et al. 1988; Kerr and Harmon, this volume) has had its greatest impact. These investigators, studying the detailed morphology of dying cells, suggested that cells die through one of only two distinct processes: necrosis and apoptosis. Apoptosis, which was observed in the embryo and adult under both physiological and pathological conditions, was considered the more significant form of cell death. The stereotyped morphology of apoptosis suggested that a common mechanism underlies the process. Moreover, the sequence of ultrastructural changes was thought to harbor valuable insights into the molecular basis of this type of cell death.

Although the original papers characterizing apoptosis risked serious oversimplification and, accordingly, generated considerable controversy, they have stimulated numerous studies of the processes by which cells die. With a few notable exceptions, these analyses have been conducted outside the nervous system. Studies of cell death in the prostate gland and thymus have generated a wealth of data and have supported the original interpretation of apoptosis as a gene-directed, morphologically stereotyped form of cell death. Molecular mechanisms have been proposed and candidate effector gene products have been identified. The goal of this review is to evaluate some of the available information on cell death in the nervous system in terms of the conceptual framework that has evolved from studies of apoptosis in other organ systems. In particular, an effort will be made to determine whether information on the process of apoptosis adds to the understanding of the mechanisms responsible for nerve cell death.

APOPTOSIS IN THE NERVOUS SYSTEM

Apoptotic cell death, as found in such tissues as thymus and prostate, is marked at the ultrastructural level by compaction of nuclear chromatin into sharply defined, uniformly dense masses that abut the nuclear envelope; condensation of the cytoplasm; convolution of the nuclear and cellular outline; fragmentation of the nucleus and of the cell; and phagocytosis of these fragments by adjacent cells wherein cellular debris is digested by lysosomes (Kerr and Harmon, this volume). The questions to be addressed herein include: (1) Does apoptosis occur in the developing nervous system? (2) Is this the only pattern of cell death? (3) Do the biochemical phenomena associated with apoptosis occurring outside the nervous system contribute to the death of nerve cells? (4) If so, is their occurrence an unequivocal marker for apoptotic nerve cell death?

The morphological characteristics of neurons dying in developing organisms were reviewed extensively by Clarke (1990). There was evidence for the occurrence of a pattern very reminiscent of apoptosis. In cells identified as undergoing degeneration, there was the appearance of dense chromatin masses within the nucleus, condensation of the nucleus with folding of the nuclear membrane, convolution of the cell membrane, loss of ribosomes from the rough endoplasmic reticulum (RER) and polysomes, and an increase in electron density of the cytoplasm associated with reduction in its volume (Fig. 1). Dying cells and cell fragments were removed by phagocytes. Clarke (1990) has referred to this pattern as Type 1 or apoptotic cell death. Examples of apoptotic cell death were found among developing thoracic spinal neurons (Pannese 1976), cervical visceromotor neurons (O'Connor and

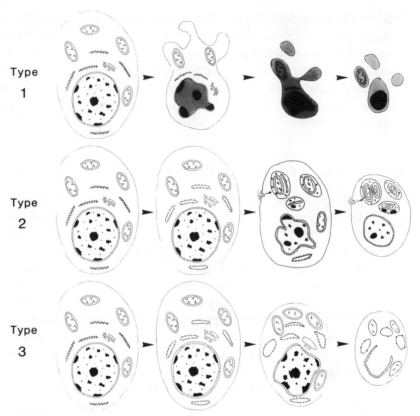

FIGURE 1 Schematic description of common patterns of nerve cell death. (Modified, with permission, from Clarke 1990.)

Wyttenbach 1974), and lateral motor neurons of the lumbar spinal cord (Chu-Wang and Oppenheim 1978) of the chick embryo. Embryonic chick ciliary ganglion neurons deprived of their normal target also demonstrated these changes (Pilar and Landmesser 1976). Type 1 cell death has also been found in mammalian neurons, having been demonstrated in the postnatal rat superior colliculus (Cunningham et al. 1982) and retina (Giordano et al. 1980). Importantly, however, morphological patterns other than Type 1 were also identified in dying neurons. Two relatively common types are represented schematically in Figure 1. Death of the second type (Type 2 or autophagic cell death) was characterized chiefly by the appearance of numerous autophagic vacuoles. Clarke (1990) noted that endocytosis may be prominent and that there may be dilation of the endoplasmic reticulum. An example of Type 2 cell death was seen in Rohon-Beard neurons in the spinal cord of *Xenopus laevis* (Lamborghini 1987), and Type 2 death was induced in neurons of the chick isthmo-optic nucleus following intraocular injection of colchicine (Hornung et al. 1989). Increased electron density of the nucleus and cytoplasm was described in the latter work. Nuclear pycnosis and chromatin clumping were also seen (Hornung et al. 1989). Type 3 or cytoplasmic cell death was seen in embryonic chick ciliary ganglion cells (Pilar and Landmesser 1976), lumbar motor neurons (Chu-Wang and Oppenheim 1978), thoracic spinal ganglion neurons (see Fig. 3B of Pannese 1976), embryonic duck trochlear nucleus motor neurons (Sohal and Weidman 1978), and postnatal rat superior colliculus (Giordano et al. 1980). This type of cell death was characterized by initial dilation of the RER without dispersion of ribosomes and by dilation of the nuclear envelope and Golgi. Mitochondria were swollen in some cases. At advanced stages of degeneration, the nucleus showed increased granularity of chromatin, and the greatly vesiculated degenerating cell was phagocytosed.

From the foregoing, it can be concluded that a morphological signature equivalent to apoptosis does mark the death of some developing neurons. What is apparent as well is that apoptosis is not the only morphological pattern to characterize developmental neuronal death. Indeed, apoptosis appears not to be the only pattern of cell death within a specific neuronal population. Chu-Wang and Oppenheim (1978) documented the occurrence of Type 1 and Type 3 cells among chick lumbar motor neurons. Both types could be seen during normal development and after the extensive (~90%) cell death induced by removal of the limb bud prior to innervation. Although the relative frequency of Type 1 and Type 3 cell deaths was not quantitated, it appeared to be the same in these two experimental contexts. Thus, the patterns of cell death created by removal of the target of developing motor neurons were the same in type and relative frequency as were seen during normally occurring degeneration. This observation raises the possibility that a similar event triggers naturally occurring and induced cell death. In keeping with earlier findings, including those of Hamburger and Levi-Montalcini (1949), this study points to the target of neurons as having a key role in regulating cell death during development of the vertebrate nervous system.

The stereotyped morphological changes that constitute apoptosis have been associated with a number of biochemical phenomena. These include an alteration in intracellular Ca^{++} (Kaiser and Edelman 1978; Connor et al. 1988; Kyprianou et al. 1988; McConkey et al. 1989), the activation of a Ca^{++}, Mg^{++}-dependent endonuclease (Wyllie 1980; Cohen and Duke 1984; Compton and Cidlowski 1986; Kyprianou et al. 1988), the degradation of nuclear DNA into oligonucleosomal fragments (Wyllie 1980; Compton and Cidlowski 1986; Kyprianou and Isaacs 1988; English et al. 1989), the requirement for RNA and protein synthesis (Stanisic et al. 1978; Cohen and Duke 1984; Wyllie et al. 1984), and the expression of specific genes including the testosterone-repressed prostate message 2 (TRPM-2) gene (Montpetit 1986). None of these biochemical events has been demonstrated in every instance of apoptosis and, accordingly, there is disagreement regarding their role in the process of apoptotic cell death. Nonetheless, in light of the suggested linkage to apoptosis, an association of any of these events with nerve cell death is noteworthy.

Evidence in support of a role for Ca^{++} in the death of nerve cells is compelling and has been the subject of recent reviews (Choi 1988; Siesjö 1988). Whereas the deleterious effect of increased cytosolic Ca^{++} is widely acknowledged, there is considerably less agreement as to which of the Ca^{++}-mediated pathways is preeminent in the death of nerve cells. In contrast to the highly specific mechanism of Ca^{++}-stimulated enzymatic degradation of DNA proposed for apoptosis, cell death in the nervous system has been attributed to the activation by Ca^{++} of numerous proteases and phospholipases as well as endonucleases (Choi 1988). The relative contribution of each of these various pathways is currently under investigation. At present, however, there is no basis for concluding that Ca^{++}-mediated neuronal cell death is invariably apoptotic in nature.

One important and, in some sense, counterintuitive concept that has evolved through studies of apoptosis is that cell death can result not only from an external insult to a cell causing the cessation of metabolic function, but also from an endogenous, gene-directed process requiring RNA and protein synthesis. Recent studies of the fate of nerve cells deprived of trophic factor support have provided convincing evidence that naturally occurring cell death in the vertebrate nervous system is, indeed, an active process (Martin and Johnson, this volume). Martin et al. (1988) demonstrated that the death of trophic-factor-deprived rat sympathetic neurons in vitro can be prevented by inhibitors of RNA and protein synthesis. Scott and Davies (1990) extended this observation to include three other trophic-factor-dependent neuronal populations: trigeminal mesencephalic neurons, dorsomedial trigeminal ganglion neurons, and ciliary ganglion neurons. In an in vivo analysis of naturally occurring and induced neuronal cell deaths in the chick embryo, Oppenheim et al. (1990) showed that the deaths of trophic-factor-dependent motoneurons and dorsal root ganglion cells require RNA and protein synthesis. The suggested interpretation of each of these results was that neurotrophic factors promote the survival of dependent neuronal populations through the inhibition of an endogenous, active cell death program (Martin et al. 1988; Oppenheim et al. 1990; Scott and Davies 1990). Although, as pointed out by Scott and Davies (1990), one cannot formally exclude the possibility that RNA and protein synthesis inhibitors merely reduce the dependence of neuronal populations on trophic factor support, the concept that these factors block an active cell death program is currently favored.

Most of the neuronal populations used in the above noted studies on the biosynthetic requirements for cell death have been evaluated at the ultrastructural level. There are detailed descriptions of the deaths of immature sympathetic neurons (Martin et al. 1988), chick ciliary ganglion neurons (Pilar and Landmesser 1976), chick motoneurons (Chu-Wang and Oppenheim 1978), and chick dorsal root ganglion neurons (Pannese 1976). As summarized above, apoptosis is not the only morphological pattern of cell death to occur within these neuronal populations, and there is nothing to suggest that only the apoptotic deaths are prevented by inhibition of biosynthesis. It appears, then, that the requirement for RNA and protein synthesis is not restricted to the apoptotic form of neuronal cell death. Stated another way, active cell death in the developing nervous system is not necessarily apoptotic cell death.

The view that apoptosis is an endogenous, active process predicts the existence of gene products that execute the death program. Efforts to identify these gene products have resulted in a number of candidate molecules and are reviewed elsewhere (Buttyan, this volume). Although there is insufficient evidence to prove that any of these molecules has an integral role in apoptotic cell death, one protein, TRPM-2, deserves special mention because of the interest it has generated and its suggested link to neurodegenerative disease. In studies of the involuting rat prostate gland, where the apoptotic pattern of cell death is clearly evident, selective induction of the TRPM-2 gene has been demonstrated (Montpetit et al. 1986; Léger et al. 1987). The molecule, also called sulfated glycoprotein 2 (SGP-2), is expressed in other mammalian tissues undergoing regression (Buttyan et al. 1989). The role, if any, of TRPM-2 in apoptosis is not known, but it has been suggested, on the basis of the protein's ability to inhibit complement-dependent cytolysis, that TRPM-2 may be synthesized by dying cells to protect surrounding tissue (Buttyan, this volume). Of particular interest is the fact that the expression of TRPM-2 or a putative human homolog has been detected in degenerating brain tissue. Duguid et al. (1989) reported elevated levels of TRPM-2 RNA in the brains of scrapie-infected hamsters, as well as in the hippocampus of patients with Alzheimer's disease or Pick's disease. Similarly, May et al. (1990) detected an increase in the expression of TRPM-2 in rat hippocampus following selective lesions that mimic aspects of Alzheimer's disease pathology and a twofold elevation of a putative human homolog of TRPM-2 in Alzheimer's disease brain. However, none of these neurodegenerative conditions has been demonstrated to result from apoptotic cell death, and any suggested link between TRPM-2 expression and apoptosis in the nervous system remains speculation.

The search for gene products that contribute to apoptotic cell death continues, primarily through studies of regressing mammalian tissues in which the classic changes of apoptosis have been documented. However, one of the most promising efforts to identify so-called "cell death genes" has used the small free-living nematode *Caenorhabditis elegans* (Horvitz and Chalfie 1991). A programmed cell death pathway has been determined through genetic analysis (Ellis and Horvitz 1986), and genes that comprise the pathway have been identified and, in some cases, cloned and sequenced (Yuan and Horvitz 1990). It was suggested that the dying nematode cells, most of which are neurons, demonstrate a morphology that resembles apoptosis, although atypical ultrastructural changes were noted (Robertson and Thompson 1982). This type of cell death, which is distinguished both morphologically and mechanistically from a second pattern in *C. elegans* (Chalfie and Wolinsky 1990), occurs during the normal development of the organism and in at least one neurodegenerative genetic disorder (Ellis and Horvitz 1986; Horvitz and Chalfie 1991). Although a detailed discussion of programmed cell death in *C. elegans* is beyond the scope of this review, a brief overview of the cell death pathway is instructive. The genes in the pathway have been characterized on the basis of function (Horvitz and Chalfie 1991); there are genes that determine which cells will die, genes that cause the death of cells, genes involved in the engulfment of dying cells by their neighbors, and genes responsible for the degradation of residual cellular debris. The actual "killer genes" are under intense investigation. They appear to act cell autonomously; that is, within the cells they kill (Yuan and Horvitz 1990). Preliminary sequence information suggests that one encodes a Ca^{++}-binding protein, and this protein could function as a Ca^{++}-dependent "deathase" (Horvitz and Chalfie 1991). The potential of this model system for the analysis of genetic mechanisms of cell death is enormous, although it remains to be determined whether pathways of cell death in *C. elegans* have any relationship to apoptotic cell death in the vertebrate nervous system.

What, then, can be concluded regarding apoptosis in the nervous system? The morphological data confirm that a stereotyped pattern of ultrastructural changes consistent with apoptosis does occur as selected nerve cells die during development. However, other discrete patterns of cell death also occur, often under the same conditions, indicating that apoptotic cell death is but one of a limited number of patterns of cell death in the nervous system. The factors that trigger apoptosis in the nervous system are not known, nor are the gene products that execute the apoptotic cell death program. Even less is known about the other patterns of nerve cell death described above as cytoplasmic or autophagic. Do they represent distinct cell death pathways or are they merely different morphological expressions of the same basic pathway? Clearly, these and related issues on the patterns of cell death in the nervous system cannot be resolved on the basis of the available data.

PROGRAMMED CELL DEATH IN THE NERVOUS SYSTEM

Despite many unanswered questions regarding nerve cell death, the studies cited above provide clear support for one concept worthy of emphasis, namely, that cell death in the developing nervous system is an endogenous gene-directed process. Accordingly, it seems appropriate to end this chapter with a discussion that extends beyond apoptosis, per se, to address the broader issue of programmed cell death in the nervous system and its role not only in development, but in aging and disease as well.

The term programmed cell death has evolved through a number of uses (Lockshin and Beaulaton 1974) but is now applied with a degree of uniformity. In his review of cell death in embryonic systems, Saunders (1966) noted that the regressive events that occur during development are "programmed to occur in a highly predictable manner" and are likely to proceed under genetic control. Lockshin and Beaulaton (1974) observed that cells dying in nonpathological situations do so in a manner that "suggests that the intracellular milieu, as well as the extracellular, is under control and that the cells, although beating a retreat, are doing so in an organized and physiological manner." These early descriptions emphasized properties that are now closely associated with the term programmed cell death—physiological, genetically controlled, organized, and predictable.

This concept of programmed cell death was evident in some of the now classic reviews of cell death in the developing nervous system. Hamburger and Oppenheim (1982) and Cowan et al. (1984) described two types of physiological nerve cell death. Programmed cell death was said to occur in certain invertebrate species in which the neurons that died during development were predestined to do so by their lineage. These neurons, which could be identified and followed throughout their lives, had a fate that was invariant. This type of cell death was distinguished from so-called naturally occurring neuronal death that was observed in the vertebrate nervous system. The latter resulted when neurons, in competition with sister cells for a limited supply of trophic molecules, were deprived of the requisite support. This form of cell death was judged to be "probabilistic or stochastic in nature" (Hamburger and Oppenheim 1982) because there were no apparent indicators as to which of the neurons would fail in their competition for trophic support, and it was concluded that no individual cell within the population was predestined to die. Regressive events in the developing vertebrate nervous system were viewed as necessary adjustments in the size of neuronal populations to match the functional requirements of their target fields (Cunningham 1982; Cowan et al. 1984).

The features that distinguished the "programmed cell death" of invertebrates from the "naturally occurring cell death" of vertebrates have received considerable attention (Hamburger and Oppenheim 1982; Cowan et al. 1984; Truman 1984; Oppenheim 1985). However, as interest has shifted in recent years from the factors that trigger nerve cell death to the actual mechanisms by which nerve cells die, the distinction between the types of physiological cell death that occur in the nervous system has blurred. The recent proposal (Martin et al. 1988) that growth factor deprivation induces naturally occurring neuronal death in vertebrates through the release of an active, gene-directed death program has forced a reevaluation of the terminology used to describe physiological nerve cell death. The naturally occurring death of developing vertebrate neurons is now described as "programmed" (Oppenheim et al. 1990), and it has even been suggested that physiological cell death in vertebrates results from the action of gene products that are functionally and structurally related to those that mediate invertebrate programmed cell death (Yuan and Horvitz 1990).

The point of this discussion is not that a single mechanism underlies all forms of physiological nerve cell death. The suggestion that apoptosis constitutes the only form of controlled cell death is not supported by the data. As indicated above, apoptosis is but one of a few morphological patterns of cell death observed in the developing nervous system. The mechanisms that mediate these different patterns of cell death have not been determined, but there is no reason to conclude that the patterns all result from the expression of a single genetic program. A review of the available information, however, does allow for a few provocative yet reasonable speculations: (1) There are only a limited number of genetic programs that lead to active, controlled cell death in the developing nervous system; (2) elements of these programs are conserved across large evolutionary distances and are utilized by both invertebrates and vertebrates; and (3) the expression of these programs is not restricted to the period of development, but can occur in aging and in some forms of disease.

With regard to the last point, Saunders (1966) was among the first to stress that genetic mutation can result in the aberrant expression of otherwise normal regressive phenomena leading to pathology. One important contribution of Wyllie and Kerr and their collaborators in their effort to characterize apoptosis was the view that this form of active, controlled cell death occurs in disease as well as in development (Kerr et al. 1972). In an insightful review, Umansky (1982) suggested that a cell death program with an essential role in the development of multicellular eukaryotes is also active in carcinogenesis and aging. Most recently, Horvitz and Chalfie (1991) have argued that an analysis of the genetic and molecular mechanisms that underlie nerve cell death in the invertebrate *C. elegans* may yield valuable insight into the pathologies of human neurodegenerative disease, stroke, and trauma to the central nervous system. These investigators and their colleagues not only have generated a comprehensive data base in support of the concept of genetic cell death pathways, but also have written convincingly on the potential of their approach to lead to innovation in the treatment of human disease (Chalfie and Wolinsky 1990; Horvitz and Chalfie 1991). If, in fact, the aberrant expression of a limited number of genetic cell death programs underlies selected disorders of the human nervous system, then insight into the function of the gene products that execute these programs could point to treatment strategies directed at the inhibition of a deleterious molecule active in a final common pathway to nerve cell death. Although the factors that trigger these cell death programs may be many and varied, the number of gene products that are actually responsible for the death of the cell and that must be evaluated as therapeutic targets may be manageably small. For those committed to the practical application of basic scientific discovery, this possibility constitutes the great promise of the growing effort to decipher the mechanisms by which nerve cells die.

ACKNOWLEDGMENTS

We thank Stephen DeArmond and Richard Davis for helpful discussions of the morphology of cell death, Peter Clarke for use of his illustration depicting patterns of cell death, and Marilyn Stubblebine for expert assistance in preparing this review. W.C.M. is supported by National Institutes of Health grants NS-24054 (NINDS) and AG-08938 (NIA).

REFERENCES

Buttyan, R., C.A. Olsson, J. Pintar, C. Chang, M. Bandyk, P. Ng, and I.S. Sawczuk. 1989. Induction of the TRPM-2 gene in cells undergoing programmed death. *Mol. Cell. Biol.* **9:** 3473.

Chalfie, M. and E. Wolinsky. 1990. The identification and suppression of inherited neurodegeneration in *Caenorhabditis elegans. Nature* **345:** 410.

Choi, D.W. 1988. Glutamate neurotoxicity and diseases of the nervous system. *Neuron* **1:** 623.

Chu-Wang, I. and R.W. Oppenheim. 1978. Cell death of motoneurons in the chick embryo spinal cord. I. A light and electron microscopic study of naturally occurring and induced cell loss during development. *J. Comp. Neurol.* **177:** 33.

Clarke, P.G.H. 1990. Developmental cell death: Morphological diversity and multiple mechanisms. *Anat. Embryol.* **181:** 195.

Cohen, J.J. and R.C. Duke. 1984. Glucocorticoid activation of a calcium-dependent endonuclease in thymocyte nuclei leads to cell death. *J. Immunol.* **132:** 38.

Collin, R. 1906. Recherches cytologiques sur le développement de la cellule nerveuse. *Le Nevraxe* **8:** 181.

Compton, M.M. and J.A. Cidlowski. 1986. Rapid *in vivo* effects of glucocorticoids on the integrity of rat lymphocyte genomic deoxyribonucleic acid. *Endocrinology* **118:** 38.

Connor, J., I.S. Sawczuk, M.C. Benson, P. Tomashefsky, K.M. O'Toole, C.A. Olsson, and R. Buttyan. 1988. Calcium channel antagonists delay regression of androgen-dependent tissues and suppress gene activity associated with cell death. *Prostate* **13:** 119.

Cowan, W.M., J.W. Fawcett, D.D.M. O'Leary, and B.B. Stanfield. 1984. Regressive events in neurogenesis. *Science* **225:** 1258.

Cunningham, T.J. 1982. Naturally occurring neuron death and its regulation by developing neural pathways. *Int. Rev. Cytol.* **74:** 163.

Cunningham, T.J., I.M. Mohler, and D.L. Giordano. 1982. Naturally occurring neuron death in the ganglion cell layer of the neonatal rat: Morphology and evidence for regional correspondence with neuron death in superior colliculus. *Dev. Brain Res.* **2:** 203.

Duguid, J.R., C.W. Bohmont, N. Liu, and W.W. Tourtellotte. 1989. Changes in brain gene expression shared by scrapie and Alzheimer disease. *Proc. Natl. Acad. Sci.* **86:** 7260.

Ellis, H.M. and H.R. Horvitz. 1986. Genetic control of programmed cell death in the nematode *C. elegans*. *Cell* **44:** 817.

English, H.F., N. Kyprianou, and J.T. Isaacs. 1989. Relationship between DNA fragmentation and apoptosis in the programmed cell death in the rat prostate following castration. *Prostate* **15:** 233.

Giordano, D.L., M. Murray, and T.J. Cunningham. 1980. Naturally occurring neuron death in the optic layers of the superior colliculus of the postnatal rat. *J. Neurocytol.* **9:** 603.

Glücksmann, A. 1951. Cell deaths in normal vertebrate ontogeny. *Biol. Rev.* **26:** 59.

Hamburger, V. and R. Levi-Montalcini. 1949. Proliferation, differentiation and degeneration in the spinal ganglia of the chick embryo under normal and experimental conditions. *J. Exp. Zool.* **111:** 457.

Hamburger, V. and R.W. Oppenheim. 1982. Naturally occurring neuronal death in vertebrates. *Neurosci. Comment.* **1:** 39.

Hornung, J.P., H. Koppel, and P.G.H. Clarke. 1989. Endocytosis and autophagy in dying neurons: An ultrastructural study in chick embryos. *J. Comp. Neurol.* **283:** 425.

Horvitz, H.R. and M. Chalfie. 1991. Implications of nematode neuronal cell death for human neurological disorders. In *New biological approaches to neurological disorders: Pathogenesis and treatment* (Dahlem Konferenzen, Berlin, August 5–10, 1990) (ed. D.L. Price et al.). Wiley, New York. (In press.)

Kaiser, N. and I.S. Edelman. 1978. Further studies on the role of calcium in glucocorticoid-induced lymphocytolysis. *Endocrinology* **103**: 936.

Kerr, J.F.R., A.H. Wyllie, and A.R. Currie. 1972. Apoptosis: A basic biological phenomenon with wide-ranging implications in tissue kinetics. *Br. J. Cancer* **26**:239.

Kyprianou, N. and J.T. Isaacs. 1988. Activation of programmed cell death in the rat ventral prostate after castration. *Endocrinology* **122**: 552.

Kyprianou, N., H.F. English, and J.T. Isaacs. 1988. Activation of a Ca^{2+}- Mg^{2+}-dependent endonuclease as an early event in castration-induced prostatic cell death. *Prostate* **13**: 103.

Lamborghini, J.E. 1987. Disappearance of Rohon-Beard neurons from the spinal cord of larval *Xenopus laevis*. *J. Comp. Neurol.* **264**: 47.

Léger, J.G., M.L. Montpetit, and M.P. Tenniswood. 1987. Characterization and cloning of androgen-repressed mRNAs from rat ventral prostate. *Biochem. Biophys. Res. Comm.* **147**: 196.

Lockshin, R.A. and J. Beaulaton. 1974. Minireview. Programmed cell death. *Life Sci.* **15**: 1549.

Martin, D.P., R.E. Schmidt, P.S. Di Stefano, O.H. Lowry, J.G. Carter, and E.M. Johnson, Jr. 1988. Inhibitors of protein synthesis and RNA synthesis prevent neuronal death caused by nerve growth factor deprivation. *J. Cell Biol.* **106**: 829.

May, P.C., M. Lampert-Etchells, S.A. Johnson, J. Poirier, J.N. Masters, and C.E. Finch. 1990. Dynamics of gene expression for a hippocampal glycoprotein elevated in Alzheimer's disease and in response to experimental lesions in rat. *Neuron* **5**: 831.

McConkey, D.J., P. Nicotera, P. Hartzell, G. Bellomo, A.H. Wyllie, and S. Orrenius. 1989. Glucocorticoids activate a suicide process in thymocytes through an elevation of cytosolic Ca^{2+} concentration. *Arch. Biochem. Biophys.* **269**: 365.

Montpetit, M.L., K.R. Lawless, and M. Tenniswood. 1986. Androgen-repressed messages in the rat ventral prostate. *Prostate* **8**: 25.

O'Connor, T.M. and C.R. Wyttenbach. 1974. Cell death in the embryonic chick spinal cord. *J. Cell Biol.* **60**: 448.

Oppenheim, R.W. 1985. Naturally occurring cell death during neural development. *Trends Neurosci.* **8**: 487.

Oppenheim, R.W., D. Prevette, M. Tytell, and S. Homma. 1990. Naturally occurring and induced neuronal death in the chick embryo *in vivo* requires protein and RNA synthesis: Evidence for the role of cell death genes. *Dev. Biol.* **138**: 104.

Pannese, E. 1976. An electron microscopic study of cell degeneration in chick embryo spinal ganglia. *Neuropathol. Appl. Neurobiol.* **2**: 247.

Pilar, G. and L. Landmesser. 1976. Ultrastructural differences during embryonic cell death in normal and peripherally deprived ciliary ganglia. *J. Cell Biol.* **68:** 339.

Robertson, A.M.G. and J.N. Thomson. 1982. Morphology of programmed cell death in the ventral nerve cord of *Caenorhabditis elegans* larvae. *J. Embryol. Exp. Morphol.* **67:** 89.

Saunders, J.W., Jr. 1966. Death in embryonic systems. Death of cells is the usual accompaniment of embryonic growth and differentiation. *Science* **154:** 604.

Scott, S.A. and A.M. Davies. 1990. Inhibition of protein synthesis prevents cell death in sensory and parasympathetic neurons deprived of neurotrophic factor *in vitro. J. Neurobiol.* **21:** 630.

Siesjö, B.K. 1988. Historical overview. Calcium, ischemia, and death of brain cells. *Ann. N.Y. Acad. Sci.* **522:** 638.

Sohal, G.S., and T.A. Weidman. 1978. Ultrastructural sequence of embryonic cell death in normal and peripherally deprived trochlear nucleus. *Exp. Neurol.* **61:** 53.

Stanisic, T., R. Sadlowski, C. Lee, and J.T. Grayhack. 1978. Partial inhibition of castration induced ventral prostate regression with actinomycin D and cycloheximide. *Invest. Urol.* **16:** 19.

Truman, J.W. 1984. Cell death in invertebrate nervous systems. *Annu. Rev. Neurosci.* **7:** 171.

Umansky, S.R. 1982. The genetic program of cell death. Hypothesis and some applications: Transformation, carcinogenesis, ageing. *J. Theor. Biol.* **97:** 591.

Walker, N.I., B.V. Harmon, G.C. Gobé, and J.F.R. Kerr. 1988. Patterns of cell death. *Methods Achiev. Exp. Pathol.* **13:** 18.

Wyllie, A.H. 1980. Glucocorticoid-induced thymocyte apoptosis is associated with endogenous endonuclease activation. *Nature* **284:** 555.

Wyllie, A.H., J.F.R. Kerr, and A.R. Currie. 1980. Cell death: The significance of apoptosis. *Int. Rev. Cytol.* **68:** 251.

Wyllie, A.H., R.G. Morris, A.L. Smith, and D. Dunlop. 1984. Chromatin cleavage in apoptosis: Association with condensed chromatin morphology and dependence on macromolecular synthesis. *J. Pathol.* **142:** 67.

Yuan, J. and H.R. Horvitz. 1990. The *Caenorhabditis elegans* genes ced-3 and ced-4 act cell autonomously to cause programmed cell death. *Dev. Biol.* **138:** 33.

Apoptosis: A Program for Death or Survival?

L.D. Tomei
Arthur G. James Cancer Hospital and Research Institute
The Ohio State University
Columbus, Ohio 43210

In this final chapter. I attempt to put some of our current thoughts into perspective and consider the experimental evidence gathered in our laboratory using the C3H-10T1/2 embryonic fibroblast and how that evidence has influenced our view of gene-directed cell death. In his discussion of the philosophy of science, Ackermann (1970) describes hypothesis as a set of statements along with all the logical consequences of that set of statements. Description of cell death, without regard to the elegance of detail with which it may be described, has limited value, since generally little insight is provided into possible means to control death short of eliminating the cytotoxic stimulus. Kerr et al. (1972) introduced the concept that cell death, marked by specific morphological changes, represented a process that could be studied in much the same way mitotic proliferation was. As such, apoptosis became an integral part of an empirical hypothesis; successful prediction of experimental observation is the ultimate purpose of this concept, as it is with all scientific hypotheses.

Apoptosis has been studied as a normal physiological process of death that occurs in embryonic development (Furtwängler et al. 1985), normal functioning in the immune system (Cohen et al. 1985), or induction by either hormonal stimulation (Wyllie 1980) or deprivation (Kerr and Searle 1973). Our laboratory has focused on apoptosis induced by cell damage caused by irradiation, treatment with cytotoxic chemicals, or serum growth factor deprivation. Together, these modes of cytotoxic stress have been referred to as toxic insult, for lack of a better term. Apoptosis that results from these disparate signals is ultimately considered to be a common process. How these different signals may be coupled to the common pathways that produce final catastrophic metabolic failure is unknown.

Gene Expression and Cell Death

Apoptosis suggests that death is a specific goal of a physiological process and not just a consequence of secondary factors. Apoptotic death of a cell can be considered as the purposeful result of the new expression of specific genes, the products of which produce catastrophic biochemical damage and metabolic failure. However, a second type of gene-directed cell death is conceivable. A specific cell type may differentiate in such a manner as to have a full or partial complement of apoptosis gene products that are blocked by limiting factors such as Ca^{++} availability. Changes in such modulating factors can then be responsible for initiation of apoptosis that is not dependent on de novo synthesis of either RNA or protein. The death of such cells would still conform to the stereotypic process of apoptosis in all general characteristics and would be considered gene-directed. We speculate that this may be the case for cells in the immune system where rapid cell death is an essential part of the process of immunity, whereas in other cell types, such as basal crypt epithelial cells of the intestine, the role of apoptosis may be quite different in the turnover of the gut mucosa and may, therefore, be under different temporal and spatial control. Given this view of apoptosis, we believe that it would not be productive to embark on the development of complex nomenclature, since this would tend to place descriptive accuracy before development of an empirically verifiable hypothesis. It may even be preferable to reduce the number of descriptive terms applied to cell death. In this view, commonly used terms such as "programmed" would not be synonymous with apoptosis and might even imply a molecular mechanism where none is yet known. Nevertheless, the term *programmed cell death* will likely remain in wide use, although perhaps it should be restricted to general descriptive purposes (Alles et al. 1991).

Hypothetical Apoptosis Genes

If apoptosis represents a complex cascade of metabolic events, as most evidence would suggest, the nature of the hypothetical apoptosis genes can be considered to constitute a modulated system. When considering apoptosis as a modulated system similar to mitosis, it is clear that several distinct functions can be attributed to the gene products. Degradation of genomic DNA (see Arends et al. 1990), alteration of cell-surface receptors (Savill et al. 1990), and increased transglutaminase activity (see Fesus and Thomazy 1988) each reflect degradative processes related to breakdown of genomic DNA and finally phagocytosis and deletion, processes that are initiated by cytotoxic stress. However, many cells in a target tissue sustain molecular damage under such stress, yet not all cells die as a consequence. As stated earlier (see Tomei and Cope, this volume), molecular damage directly resulting from toxic insult should not be presumed to be the cause of death. Conversely, survival of cells should not be attributed to a lack of molecular damage without further analyses. In most instances, the survival of cells must be accounted for as well as their death, if all cells have received the same toxic insult, and insight into the modulation of apoptosis may thus be provided by examination of surviving cells. Therefore, apoptosis genes may also be responsible for blockade of cell death. A clue to the nature of the hypothetical genes may be found in recent evidence regarding the role of oncogenes. Hockenberry et al. (1990) recently presented evidence that the *bcl-2* gene is associated with inhibition of apoptosis. This is in accord with other data indicating that expression of *ras* oncogenes increases cell survival to ionizing radiation at doses demonstrated to induce apoptosis (Sklar 1988). Other workers have shown that DNA damage by ionizing radiation markedly elevates expression of c-*myc* (Sullivan and Willis 1989) and that DNA-damaging agents induce expression of *fos* RNA (Hollander and Fornace 1989).

Two general lines of thought are briefly presented here based on the idea that apoptosis involves expression of gene products which lead to cellular degradation as well as inhibition of the process as in a feedback control system. The first analyzes the early degradative events within the apoptotic cell, whereas the second examines the evidence that modulation of apoptosis is closely linked to cell-cycle control through cell:cell interactions.

DNA Degradation

Internucleosomal fragmentation. An intriguing molecular marker associated with apoptosis, and certainly the most widely investigated, is endonucleolytic fragmentation of nuclear DNA. A substantial number of early reports in the literature suggested that endonucleolytic chromatin degradation accompanied cell death following irradiation (Cole and Ellis 1957; Kurnick et al. 1959; Swingle and Cole 1967). However, Wyllie (1980) provided a critical link between the changes in nuclear ultrastructure and the molecular biology of dying cells. In their studies of glucocorticoid-induced thymocyte death, these authors reported that apoptotic cells fragmented their DNA into double-stranded units that were multiples of the nucleosomal organization. The internucleosomal fragmentation could be demonstrated by the formation of a ladder banding pattern in neutral agarose gel electrophoresis separation of nuclear DNA from apoptotic cells. Nikonova et al. (1982) presented early compelling evidence of internucleosomal DNA fragmentation in thymocytes treated with either glucocorticoids or γ irradiation that was blocked by cycloheximide. These findings were confirmed and extended by several other investigators soon thereafter (see Kerr and Harmon; Duke; and Umansky, all this volume). Subsequently, Wyllie et al. (1984) confirmed that nuclear ultrastructural changes first associated with apoptosis were accompanied by internucleosomal cleavage, which was dependent on RNA and protein synthesis and the presence of Ca^{++}. This phenomenon has been reported in numerous subsequent publications by investigators who have studied apoptosis in several other cell types in addition to thymocytes. In all, these observations have led to the concept that apoptosis is marked by the cleavage of DNA in the internucleosomal linker region, a process presumably dependent on gene expression and mediated by a Ca^{++}-dependent endonuclease(s) (see Arends et al. 1990).

As is often the case in biology, not all experimental evidence is in accord with this view. Several groups have reported that apoptosis can be initiated in the absence of new RNA or protein synthesis; generally, these findings have been based on the generation of the characteristic DNA fragmentation ladder in neutral electrophoretic gels as evidence of apoptosis. Al-nemri and Litwack (1990) observed internucleosomal cleavage following treatment of CEM cells (a lymphocyte cell line) with glucocorticoid or novobiocin in the absence of de novo protein synthesis. They also found no evidence that this process was dependent on Ca^{++}, which led these authors to conclude that apoptosis involves a constitutive non-Ca^{++}-dependent endonuclease and does not necessarily involve initiation of new gene expression. As mentioned above, such evidence is not necessarily inconsistent with the concept of apoptosis as gene-directed cell death, since hormonal stimulation of apoptosis in immune cells may represent differentiation-dependent changes in a basic physiological process. However, it is not clear how these observations can be reconciled with those reported by other investigators suggesting a requirement for protein synthesis and activation of a Ca^{++}-dependent endonuclease.

A number of investigators have found that cell death following a variety of toxic stress modes does not lead to inter-nucleosomal double-stranded fragmentation (Duke et al. 1988; see also Duke, this volume). Failure to observe fragmentation is difficult to explain, since the mechanism of thymocyte DNA fragmentation is not yet known. However, our laboratory has concluded that it is not likely that apoptosis is uniquely associated with internucleosomal cleavage. Apoptotic DNA degradation can also be demonstrated using linear sucrose gradient sedimentation analysis (Kanter et al. 1982, 1984; Tomei et al. 1985, 1986). Kanter and Schwartz (1980) originally described the phenomenon of postrepair DNA damage in X-irradiated human tumor cells.

Apoptotic degradation and higher-order chromatin organization. At about the same time that Kerr et al. (1972) made their observations regarding the phenomenon of apoptosis, several other groups began to notice that cells treated with toxic levels of a number of different agents exhibited remarkably similar patterns of DNA fragmentation, regardless of the presumed nature of the cell damage. Williamson (1970) reported that internucleosomal fragmentation preceded cell death in cultures of embryonic mouse liver cells, but the significance of this observation was not clear at that time. In 1974, Williams et al. described similar patterns of DNA degradation after exposure of cells to specific cytotoxic antibody or to detergent, or after low-energy β irradiation by high-specific-activity tritiated thymidine ([^{3}H]dThd) incorporated into cellular DNA. These authors concluded that widely differing mechanisms of cellular trauma produced a common mechanism of cell death. Matyasova et al. (1979) demonstrated breakdown of the DNA of murine lymphoid cells after exposure to X irradiation and nitrogen mustard-type alkylating agents. The regular character of the breakdown species was similar to that observed after micrococcal nuclease digestion of isolated chromatin. The DNA fragmentation pattern originally described by Williams et al. (1974) was reproduced in our laboratories using C3H-10T1/2 cells, which were either exposed to high-specific-activity [^{3}H]dThd during exponential proliferation or deprived of serum growth factors (Tomei et al. 1985, 1988). The results of these experiments indicated that the common pattern of DNA fragmentation observed by other workers using toxic cell treatments was also observed in a distinct fraction of cells following serum deprivation. Furthermore, the DNA fragmentation was found to occur prior to the physical disintegration of the cells, which morphologically corresponded to the time of cell death.

Recently, we have examined the fate of DNA through analyses in both sedimentation gradients and gel electrophoresis under both neutral and alkaline conditions. Earlier, both methodologies had led to similar conclusions regarding apoptosis with respect to inhibition by tumor promoters as well as cycloheximide. However, we wished to consider further the nature of the linkage between chromatin changes as revealed by electron microscopy, and cleavage of DNA in the nucleosomal linker region specifically in cells where apoptosis was not associated with internucleosomal double-stranded cleavage.

As shown in Figure 1, similar patterns of DNA damage appear in cells after X irradiation and exposure to DNA-reactive

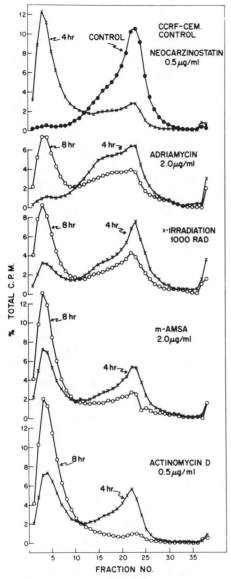

FIGURE 1 Effect of DNA reactive agents on sedimentation of CCRF-CEM leukemia cell DNA in alkaline sucrose gradients. Prelabeled ([2-^{14}C]dThd) cells were exposed to agents for 2 hr, then washed and resuspended in fresh growth medium (see Kanter et al. 1984). Lower weight DNA was associated with nonadherent, viable cells (based on trypan blue dye exclusion) and apoptosis. Unlike thymocytes or C3H-10T1/2 cells, these cells respond to 0.5 µg/ml actinomycin D by apparently initiating apoptotic DNA fragmentation.

anticancer agents. As our laboratory previously reported (Kanter et al. 1984), postrepair DNA fragment sizes were approximately normally distributed within the linear sucrose gradient for all cell lines examined. The frequency distribution of DNA strand length would then be considered logarithmic-normal (i.e., lognormal). Therefore, we infer from these data that the alkaline-sensitive sites along the length of the DNA molecule do not occur at regular intervals but rather at random locations with respect to strand length. Unlike the uniform frequency of internucleosomal linker sequences, the level of structural organization reflected in the nonrandom distribution of fragment sizes is probably of higher order and may be expected to have substantially more strand length variability between sites.

The sedimentation profiles using both cytotoxic drugs and X irradiation, as well as those reported earlier using growth factor deprivation and high-specific-activity [3H]dThd incorporation showed only two independently distributed populations of fragment sizes; these two size distributions have been referred to as Mn_1 and Mn_2 (Kanter and Schwartz 1980). The modal fragment size in each respective distribution is approximately 6×10^8 D for Mn_1, and 6×10^5 D for Mn_2. A minimum in the overall distribution was found to be in the 10^7 D to 10^8 D range, suggesting that the DNA damage resulting indirectly from a variety of cellular insults is marked by a reduction from units consisting of several thousand nucleosomes to units equivalent to between 7 and approximately 20 nucleosomes. On the basis of the conclusion that the fragment size distribution was lognormal, we proposed that the stereotypic alkaline sedimentation profiles reflect breakdown of genomic DNA at a higher level of organization, presumably associated with mechanisms of chromatin compaction.

It is known that there is a substantial degree of strand length variation within the 25–30-nm "superbead" structures of the DNA molecule (for review, see Nover and Reinbothe 1982). Furthermore, these superbead structures have been proposed to have approximately 5–7 nucleosomes per turn, determined principally by H1 histone interactions within the helical superstructure. This frequency agrees well with the observed apoptotic DNA fragment distribution Mn_2 value of 6 x

10^5 D. The data would suggest that apoptotic chromatin degradation involves damage to structural elements in the DNA at the next higher level of organization from that of the nucleosomal linker, which would have yielded double-stranded fragments of about 180 bp in length and a range of molecular masses of 1.08×10^5 D to 1.26×10^5 D. The alkaline sedimentation profiles also indicate that there is little in the range between 10^7 D and 10^8 D which separates the Mn_1 and Mn_2 distributions. The repeating structural feature of the DNA molecule that occurs at this frequency would be the DNA superbeads that comprise the next higher level of organization from that of nucleosomes.

Although modification of nucleoprotein interactions could lead to accessibility to double-strand cleavage sites by endonucleases, no direct evidence has been presented in support of such a mechanism. Umansky et al. (1981) concluded that there was no evidence that proteolytic enzymes were involved in chromatin changes observed in apoptotic rat thymocytes, whereas Kyprianou and Isaacs (1988) later arrived at similar conclusions in studies of rat ventral prostate cells after castration. Such changes may have accounted for increased accessibility of sites to constitutive endonucleases. As discussed later in this section, the findings of Arends et al. (1990) differ, since these authors noted H1 histone depletion in fragmented DNA, but it is as yet not clear whether this applies to larger oligonucleosomal fragments. This implies that higher-order DNA degradation may be determined by alterations in histone interactions, particularly those of the H1 histones.

Generally, our observations have led us to conclude that apoptosis involves modification of chromatin structure, which results in a breakdown of the supercoiling organization and formation of individual superbeads. These structures have a lognormal distribution of strand lengths and a modal size of approximately 7 nucleosomes. Many cell types, including cell lines such as HeLa and C3H-10T1/2, exhibit this type of DNA degradation in apoptosis with no evidence of double-stranded internucleosomal cleavage as found in glucocorticoid-treated thymocytes. As illustrated in Figure 2, these data also suggest that apoptosis involves degradation of supercoiled DNA

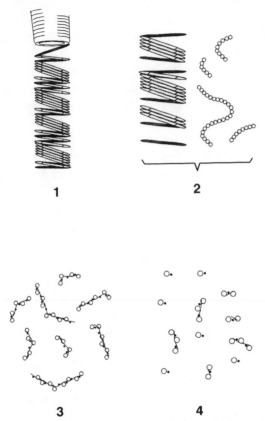

FIGURE 2 Schematic representation of supercoiled DNA and the proposed degradation cascade in apoptotic cells: (1) supercoiled DNA molecule with 7–10 nucleosomes per helical turn; (2) double-stranded fragmentation giving rise to "superbead" units containing multiples of 7–10 nucleosomes; (3) further degradation of nucleosomal compaction, which may provide access to internucleosomal linker regions associated with H1 histone sites; and (4) final cleavage of internucleosomal linker strands to ultimately yield mononucleosomes.

through double-strand breakage, which occurs at intervals equivalent to single superhelical turns of approximately 6–7 nucleosomes. Evidence from parallel neutral sedimentation analyses (Kanter and Schwartz 1980) indicated that a double-strand break occurs at randomly spaced sites along the supercoiled structure, which would then yield the lognormal distri-

bution of strand lengths between 6 x 10^5 D up to approximately 10^7 D in weight. As with internucleosomal breaks, the nature of the endonuclease responsible for degradation at this
level of structural organization is not known. However, Promwichit et al. (1982) reported that postrepair DNA damage as
originally described by Kanter and Schwartz (1980) (and subsequently linked with apoptosis [Tomei et al. 1988]) was accompanied by a marked suppression of DNase I inhibitor activity. It is possible that modulation of endonucleolytic activity
could be achieved through DNase I inhibition by monomeric
nuclear actin, a process dependent on increased Ca^{++}
(Hitchcock 1980; Limas 1981; Malicka-Blaszkiewicz and Roth
1981). We reported that both tumor promoters and retinoids
inhibited apoptosis in the C3H-10T1/2 cells (see Table 2),
which is consistent with the results reported by Rao (1985)
that both phorbol esters and retinoids induce actin
polymerization in human leukocytes.

Apoptosis in cells of the immune system, such as
thymocytes and lymphocyte-derived cell lines, involves internucleosomal double-strand cleavage; however, other cell types
such as HeLa and untransformed C3H-10T1/2 do not (P.M.
Kanter et al., in prep.). Our proposal that expansion of the
apoptosis model to include chromatin structural modifications
at levels above nucleosomal unit organization could resolve
some of the apparent contradictory results obtained over
recent years regarding the role of Ca^{++}-dependent and Ca^{++}-
independent endonucleases, de novo RNA and protein
synthesis, and ultimately the question of whether apoptosis is
indeed gene-directed cell death. Therefore, we considered the
question of whether there may be some DNA damage in the
nucleosomal linker region even in the absence of frank double-
strand cleavage.

Apoptosis and internucleosomal linker modification. Internucleosomal fragmentation presumably involves the synthesis
of an enzyme capable of recognizing the short linker region. As
Arends et al. (1990) have shown, apoptotic nuclear DNA consists of long oligonucleosomes rich in H1 histones and short
oligo- and mononucleosomes markedly depleted in H1
histones. On the basis of the analysis of sedimentation profiles

as presented above, it is possible that an early event in apoptosis would involve histone interactions related to the maintenance of supercoiling structures observed by electron microscopy and related to the mechanism of chromatin compaction. The data suggest that histone H1 may also be involved in endonuclease recognition of the linker sites, perhaps in conjunction with single-strand modification. It is not clear how H1 histones may be involved in chromatin degradation at the level of supercoiling. Searches for a double-strand endonuclease have not yet led to a definitive identification. Our recent experiments have led us to speculate that internucleosomal fragmentation may involve not a double-strand endonuclease, but rather a single-strand modification induced by a constitutive enzyme that recognizes stable apurinic or apyrimidinic (AP) sites. This view is based in part on recent data in which we compared neutral and alkaline gel electrophoretic patterns of DNA from apoptotic C3H-10T1/2 cells (data not shown).

Our current investigations have focused on the possibility that single-strand breaks revealed under alkaline conditions could be due to frank strand breakage, or alternatively, this may indicate the presence of stable alkaline-sensitive sites introduced in the linker region of the apoptotic cell DNA. We have proposed a modified view of internucleosomal damage in apoptosis, as shown in Figure 3. Alkaline-sensitive fragmentation can indicate the presence of AP sites in the genomic DNA. Our recent electrophoresis evidence supports the concept that the AP sites are introduced in a single strand specifically in the nucleosomal linker portion of the molecule. This type of modified DNA structure has been demonstrated in a wide variety of cell types and has been shown to be stable in many instances (Duker et al. 1982). Formation of AP sites has long been closely associated with mutagenic and carcinogenic activity of a variety of DNA-damaging agents, but no clear linkage with the biology of these processes has been established (Fortini et al. 1990; Laval et al. 1990). It is possible that AP sites may be introduced either directly by the DNA-damaging agent, or through constitutive DNA repair enzymes. Such a model for apoptosis would also predict that the formation of AP sites in the linker region, presumably in association

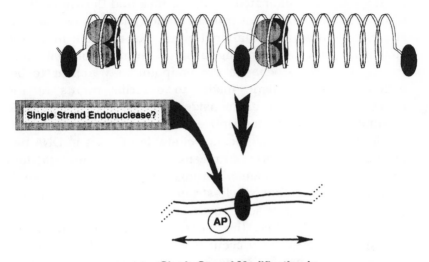

Single Strand Modification in

Internucleosomal Linker Region of DNA

FIGURE 3 Schematic diagram of internucleosomal DNA modification associated with apoptosis. A cascade of events leads to single-strand modification in the form of production of an alkaline-sensitive site in the internucleosomal linker region. Lymphocytes are then capable of proceeding to double-strand cleavage and free nucleosomal units, whereas other cell types such as the C3H-10T1/2 are arrested at the point of strand modification.

with H1 histones, would not be repaired as would similar sites in other regions of the DNA molecule. Of course, the presence of alkaline-sensitive sites in the linker region of apoptotic C3H-10T1/2 cells may reflect a defective internucleosomal cleavage mechanism in comparison with that of thymocytes.

As stated earlier, much can be learned from the study of cells that have survived apoptotic stimuli such as hormones or cytotoxic insult by chemicals or irradiation. As Elkind (1985) points out, there is a clear association between DNA damage and cell death, but there is a long-recognized, substantial disparity between the dose level of DNA-damaging agents required for cell killing and the level required to produce the characteristic DNA lesions on which the molecular models rely (see also Lutz 1990). A similar disparity between mutagenesis and DNA damage has been evident to researchers in this field for some time, as indicated by McCormick and Bertram (1982), who noted that malignant transformation may not be the consequence of a single mutational event, but rather it may be a stochastic process involving a large number of cells and events. However stochastic the endpoint may appear to be, without a deterministic model to describe events at the molecular level as we are considering here, any probabilistic treatment will remain descriptive in nature.

The physiological significance of stable AP sites in DNA has been a subject of interest in recent years among investigators searching for mechanisms of mutagenesis. The model presented here proposes that single-strand DNA modification in the nucleosomal linker region may constitute a critical early step in apoptosis. However, the model also implies that single-strand modification may occur at many sites within DNA but that these sites are not equivalent with respect to their role in the mechanism of apoptosis. It is possible that introduction of stable AP sites is an element in the signal transduction pathway that presumably leads to internucleosomal fragmentation, transcription failure, and death. However, it is also conceivable that introduction of these sites through constitutive DNA repair enzymes may be part of the degradative cascade that is initiated elsewhere. Utsumi et al. (1990) have recently proposed that the repair of sublethal DNA damage is distinct from that of lethal damage, although there is no compelling evidence that there is a difference in the molecular nature of that damage. Therefore, it is our view that lethal and sublethal damage can be distinguished on the basis of the location of the damage relative to the structural organization of the DNA molecule. Perhaps if damage occurs in the nucleosomal linker region of the DNA, this damage constitutes a signal that leads to death through apoptosis.

Alnemri and Litwack (1990) found that novobiocin was capable of initiating nucleosomal fragmentation in CEM-C7 cells without the apparent involvement of new protein synthesis. These authors suggest that novobiocin induction of nucleosomal cleavage through activation of an endogenous endonuclease was related neither to topoisomerase II nor to a Ca^{++}-dependent process. Rather, they propose the existence of a constitutive endonuclease that is activated in some manner by the novobiocin. In view of the suggestion by Utsumi et al. (1990) that novobiocin can be used to distinguish between potentially lethal damage and sublethal damage, it is possible that that distinction is related to novobiocin's ability to activate a special endonuclease that recognizes those potentially lethal sites. On the basis of our model, these sites would be expected to be found in the internucleosomal linker region in the form of stable AP sites that form part of an active recognition site for a specialized AP endonuclease. The remainder of the recognition site may be the close association with the H1 histone. Presumably, cells that are deficient in this endonuclease would tend to have transcriptional activity preserved, at least until de novo synthesis of this protein was accomplished.

In summary, these observations suggest that apoptosis involves a cascade of events with an early step in which strand modification occurs at nonrandom sites that are closely associated with the compaction mechanism of nuclear DNA. This would imply that early critical events involve proteolytic modification and double-strand cleavage at accessible sites along the supercoiled structure. The consequence would be formation of double-strand fragments consisting of multiples of the 7–10 nucleosomes. Following the loss of higher-order chromatin structural organization, single-strand modification would precede appearance of double-strand breaks in the internucleosomal linker region, a process that does not go to completion in some cells, such as the C3H-10T1/2 or HeLa. This would account for the observation that thymocytes fragment their DNA within a few hours of treatment with, for example, glucocorticoid and physically degenerate soon after that, whereas other cell types may take several days to die. The degree to which DNA tertiary structure is preserved during apoptosis and the resultant ability of the cell to support new protein synthesis may be related to the structural organization level at which degradation occurs in the nuclear DNA. Since 10T1/2 cells remain viable for several days after DNA fragmentation is observed, we speculate that there is maintenance of transcriptionally active nucleosomes even in the presence of apoptotic single-strand modification in the form of stable AP sites.

Apoptosis, Cell-cycle Control, and Intercellular Communications

After considering some of the molecular events that have been associated with apoptosis, there is yet a need to relate these intracellular processes to the physiology of cell populations. Regardless of how accurately a biochemical model may describe cellular response, the investigator is always faced with the task of considering why many cells in a culture dish or tissue sample may not appear to respond to a stimulus. When considering, for example, radiation-induced apoptosis, failure of any particular cell to die does not imply lack of damage, and we feel that simply describing survival as radioresistance is inadequate. Rather, a compelling association must be established between the molecular events and the cell physiology. Careful consideration should be given to possible modulation of cell responses through cell:cell interactions, which is a parameter recognized as important but often too difficult to account for experimentally. Here we briefly examine the evidence that cultures of C3H-10T1/2 cells express intercellular modulation of both cell-cycle control marked by G_1/S transition, and apoptosis marked by DNA fragmentation, through mechanisms related to cellular adhesion.

Kerr et al. (1972) emphasized that apoptosis complements mitosis. Together these model processes provide a more comprehensive description of cell-cycle control in normal cell turnover within tissues, and in cell populations that are either expanding or diminishing in number. However, it is readily appreciated that study of apoptosis control presents significant experimental design problems not encountered when considering cell proliferation alone. A critical design consideration is the choice of model cell systems in which to study apoptosis. Although the study of cells in situ within tissues may be preferable from a theoretical standpoint, the vast majority of experiments will be conducted using conventional cell culture techniques. No studies have been reported on the degree of variation that may exist among cells with regard to the capability to initiate apoptosis. We believe there is a substantial amount of indirect evidence that a cell's ability to initiate apoptosis is suppressed or even lost during transformation or as a consequence of establishment as stable clonally homogeneous cell lines (Tomei and Wenner 1981). Cell lines are cell populations selectively established on the basis of the cell's ability to proliferate and, notably, not its ability to die. In view of the evidence that apoptosis is an important element in the array of normal phenotypic markers, traditional criteria for successful establishment and maintenance of cell cultures based on "high viability" may be viewed with less enthusiasm. The ability to initiate apoptosis is a phenotype selected against or even completely lost in the process of establishing a cell line from primary tissue.

Our laboratory has focused most of our in vitro studies on mouse embryonic C3H-10T1/2 cells that express several critical, normal phenotypic markers, such as density-dependent G_1 arrest and sensitivity to malignant transformation. We have reported evidence of multiple G_1 substates associated with cell-cycle control mechanisms that are lost following malignant transformation by chemicals, radiation, or viruses (Tomei et al. 1981). In addition to the great deal of evidence based on proliferative control, there is an important property in the ability of cells to die under various conditions in predictable numbers. This property may be demonstrated by, for example, stable cloning efficiencies of less than 30% and modulation of saturation density with appearance of apoptotic cells as a function of serum concentration. These characteristics are lost following either spontaneous or carcinogenic transformation (Heidelberger et al. 1983). We studied cell-cycle control defined in terms of both proliferation and death using several tumor promoters such as phorbol esters and teleocidin. These agents had a profound influence on population dynamics at very low concentrations of 10^{-10} M to 10^{-8} M, which implied that the mechanism of action was intimately associated with physiological growth control mechanisms. Some of our laboratory's earliest clues to the ability of tumor promoters to inhibit apoptosis were based on observations of enhanced cloning efficiencies. Curiously, several investigators have reported that various agents produced similarly enhanced cloning efficiencies in a variety of cells in vitro. No molecular mechanisms were proposed to account for the "trophic" effects, and experimental designs often compensated for the phenomenon by seeding fewer cells in those cultures to be treated with phorbol esters in order to obtain the same number of clones as in control cultures (Kennedy et al. 1980).

Cell:cell interactions are generally recognized as important in descriptions of cell-cycle control, although not often have such spatially defined processes been measured. Cell survival has been demonstrated as having a substantial dependency on cell:cell interactions, but it is often evident from these data that survival and proliferation are not easily distinguished. However, the in vitro data accumulated over several decades have revealed that diffusible, secreted cell products exert a profound influence over cell proliferation and cell death (Revesz, 1956; Millar et al. 1978; Schorpp et al. 1984; Lynch et al. 1986).

Our laboratory began studies on the two-dimensional analysis of cell-cycle control and cell death simultaneously in the same closely interacting cell population. Early in these studies, it became evident that tumor promoters had an even more complex effect on G_1 control than anticipated. This led us to propose the G_1 "staging" protocol to kinetically define substates within a model feedback control system (Tomei et al. 1981). At that time, the output of such a system was presumably S-phase entry. As shown in Figure 4, TPA pretreatment resulted in a steady-state distribution of cells in two distinct positions within G_1 based on subsequent S-phase entry kinetics. Furthermore, cells within G_1 appeared to be distributed into multiple states, and the effects of TPA could not be described simply in terms of increased or decreased G_1/S transition as generally thought. We proposed a multistate model for phorbol ester actions on cell proliferation; however, the model was not adequate to account for the often large discrepancy between the number of cells entering S phase and the total number of cells present at the end of an experiment. Clearly, TPA was having an effect on cell survival under the culture conditions used in these experiments.

Drug effects on cell survival and cell cycle could be measured in replicate cultures by counting the number of adherent versus nonadherent serum-deprivation released (SDR) cells after serum deprivation. It was soon evident that the ability of several agents to decrease the number of SDR cells was directly related to their ability to induce trypsin-resistant

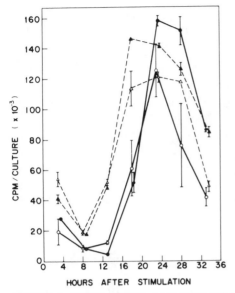

FIGURE 4 Effect of TPA pretreatment on the proliferative response of serum-deprived 10T1/2 cells. 10T1/2 cells were seeded at 10^4 cells/dish and during exponential growth (3 days), they were serum-deprived for 24 hr. The dishes were then divided into two groups and were serum-deprived for an additional 24 hr in the absence (circles) or presence (triangles) of 10^{-7} M TPA. At zero time, serum-containing medium was added to all dishes and, in addition, 10^{-7} M TPA was added to half of each of the above two groups (circle and triangle). At the indicated times, 1 μCi [^{3}H]dThd was added to each dish for 1 hr, the dishes were washed, and the cells were fixed with 1.5% PCA. The acid-insoluble material was collected and counted by liquid scintillation counting. The results are expressed as cpm/culture ± S.E. (Reprinted, with permission, from Tomei et al. 1981.)

adhesion. Since SDR cells were also linked to postrepair (i.e., apoptotic) DNA fragmentation (Kanter et al. 1984), and since agents such as TPA were also mitogenic, it was then possible to relate the cell-cycle activation effects with inhibition of cell death. The results suggested that phorbol ester stimulation of cell cycle was associated with specific changes in cell adhesion mechanisms and that inhibition of apoptotic death occurred in a distinct and separate subpopulation of cells within a single clonally homogeneous culture. Moreover, the spatial analysis of these effects also suggested that these effects were modulated by cell:cell interactions, presumably through secreted diffusible cell growth factors. Some of that evidence is briefly presented here.

Spatial analysis of G_1/S transition.   If cell-cycle activation by phorbol esters is directly linked to inhibition of apoptosis, then two relationships must exist between dose and effect. The first is simply a dose dependency usually described as logarithmic in nature and involves counting total number of cells that respond during a specific time interval. A second relationship is spatially defined and actually is a corollary of the first statement; if we examine the cell responses in small fields, or domains, of 100–300 cells (a convenient size under conventional microscopes) in a single typical monolayer of approximately 1,000,000 cells, then the number of responding cells for each domain should be related to the dose of the drug as described by the bulk responding cell count. Deviations from a population mean value should then be due to chance alone if all cells are equivalent. As shown in Figure 5, total nuclear density of C3H-10T1/2 cells was plotted against the corresponding number of radiolabeled nuclei observed in each of the field density categories. This approach revealed no relationship between the total number of cells present in any particular domain and the number of cells that had entered S phase in response to the uniform mitogenic stimulus. However, if these cultures were treated with trypsin to remove all trypsin-sensitive cells, only then was a direct relationship between nuclear density and the number of radiolabeled nuclei demonstrable. Therefore, we tentatively concluded not only that TPA was mitogenic, but also that it increased cooperative cell:cell interactions modulating G_1/S transition within the limited spatial domains.

These results implied that simple serum deprivation of a clonally homogeneous cell population led to segregation of distinct subpopulations, only one of which was linked to mitogenic response. As shown in Table 1, TPA influenced not only S-phase entry, but also the number of cells surviving. From this basic experiment, we could conclude that TPA-induced S-phase entry occurred in a stable adherent cell subpopulation, whereas enhanced cell survival was dependent on

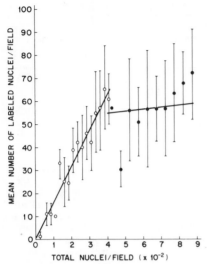

FIGURE 5 C3H-101/2 cells were deprived of serum during exponential growth by replacing growth medium with serum-free medium containing 10^{-7} M TPA. At T = 6 hr, 0.2 μCi/ml [^{3}H]dThd was added to each culture and at T = 24 hr were prepared as follows: The first group (filled circle) of replicate plates was fixed and prepared for autoradiographic analysis (-), whereas the second group (open circle) was treated with 0.25% trypsin (30 min, 37°C) to remove all trypsin-sensitive cells prior to fixation and preparation for autoradiography. Duplicate plates from each group were then analyzed by counting at least 100 microscope fields, recording both the total number of nuclei and the number of labeled nuclei in each field. Data were grouped into categories of (1–50), (51–100), ($n - [n + 49]$). The mean number of labeled nuclei was calculated for each category and the S.D. was used to estimate the distribution about each mean. A weighted least-squares regression was then calculated for both untrypsinized and trypsinized groups. (Reprinted, with permission, from Wenner et al. 1984.)

a distinctly different, unstable, nonadherent population that remained responsive for less than 48 hours after serum removal. As discussed earlier, this latter population was found to be linked with apoptosis as defined by protein synthesis-dependent specific DNA fragmentation.

Figure 6 shows the relationship between the dose-dependent TPA induction of trypsin resistance and the induction of S-phase entry. As can be seen, at 10^{-8} M TPA, more

TABLE 1   *TPA EFFECTS ON C3H-10T1/2 CELLS AFTER SERUM REDUCTION*

Period of TPA/[^{3}H]dThd addition (hr)	Cumulative nuclear labeling 10^3 labeled nuclei/cm^2		10^4 viable cells/cm^2	
	$-TPA$	$+TPA$	$-TPA$	$+TPA$
0–24	2.23 ±0.03	6.13 ±0.66	8.38	15.18
24–48	0.52 ±0.01	7.80 ±0.52	6.24	11.54
48–72	0.88 ±0.02	5.97 ±0.70	5.85	5.72

Cultures were seeded at 10^4 cells/60-mm plate and allowed to proliferate until ~10^5 cells/plate (BME + 10% fetal bovine serum). Medium was then replaced with BME containing 0.5% serum and 0.1 µM TPA + [^{3}H]dThd (1 µCi/plate) added at 0, 24, and 72 hr. After 24 hr, each plate was fixed and prepared for autoradiographic analysis. The number of labeled and nonlabeled nuclei was determined using an electronic image analysis system (Artek, Inc.) attached to a Nikon inverted phase-contrast microscope. Each field corresponded to 6×10^{-3} cm^2 and the results are expressed in terms of the mean ± S.E., where n = 100 fields in duplicate culture plates.

than 90% of the cells that entered S phase belonged to the trypsin-resistant subpopulation, which comprised less than 20% of the total population. The cells "rescued" from apoptosis were limited to those cells removed by trypsinization, suggesting that two morphologically indistinguishable cell types could be segregated by their distinct adhesion mechanisms. As seen in Figure 7, TPA inhibited appearance of apoptotic SDR cells (panel a), which was detectable at 15 hours in untreated control cultures. Simultaneously, TPA increased trypsin-resistant adhesion as soon as 3 hours after serum deprivation. However, the most rapid change in cell adhesion was marked by Ca^{++} sensitivity (panel c), which was detected within 5 minutes of serum deprivation. Dose-response curves could then be constructed using measurements of either the reduction in apoptotic SDR cells or the induction of trypsin-resistant adhesion as illustrated in Figure 8. Comparison of activities observed for a variety of agents found to induce trypsin-resistant adhesion and to suppress formation of apoptotic cells in C3H-10T1/2 cultures is presented in Table 2. It can be seen that the relative activities of these agents appeared to be related to the specific tumor promoting activity of each. For example, phorbol didecanoate (PDD), a highly active tumor promoter,

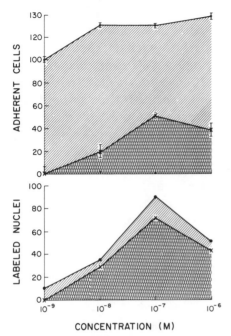

CONCENTRATION (M)

FIGURE 6 Relationship between dose-dependent phorbol ester induction of trypsin resistance (TR) (upper panel) and induction of S-phase entry (lower panel). TR response was measured by counting cells in replicate plates of C3H-10T1/2 before (filled circle) and after (x) treatment with 0.1% trypsin. The S-phase entry was determined simultaneously by counting the number of [^{3}H]dThd-labeled nuclei. Replicate cultures were treated with TPA serum-free medium containing 1 μCi/5 ml of high-specific-activity [^{3}H]dThd (60 Ci/mmole) in order to prevent repetitive cycling of responding cells during the 24-hr cumulative labeling period.

was substantially more active than the 4-*O*-methyl derivative of TPA, which is inactive as a tumor promoter. However, we also noted that some retinoids were also capable of inhibiting formation of apoptotic SDR cells as well as inducing trypsin-resistant adhesion. Therefore, the relationship between these adhesion responses and tumor promotion may be dependent on the additional ability of phorbol esters to induce S-phase entry.

Spatial factor defined. We considered the implications of TPA inhibition of apoptosis (i.e., reduction of SDR cells), stimula-

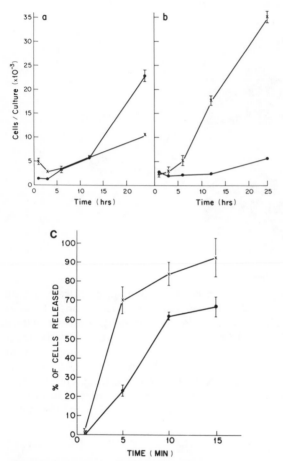

FIGURE 7 Kinetics of specific adhesion responses induced in exponentially proliferating C3H-10T1/2 cells following serum deprivation in the presence (X) and absence (filled circle) of 10^{-7} M TPA ($\pm$ S.E.M., $n =$ 6). (a) TPA reduction in the number of cells released from culture plate surface. These cells have been referred to as serum deprivation released (SDR) cells and were found to contain specifically fragmented DNA (see Kanter et al. 1984). (b) Kinetics of trypsin-resistance induction by TPA, which was determined by counting the number of adherent cells remaining after trypsinization (0.25% trypsin, 37°C, 30 min). (c) Effect of TPA on the release of cells from the culture plate surface using only Ca^{++}, Mg^{++}-free phosphate-buffered saline (pH 7.4) containing 0.02% EDTA.

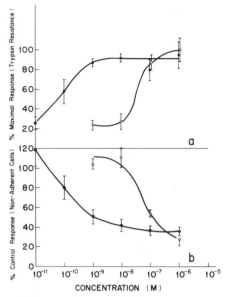

FIGURE 8 Comparison of dose-response curves obtained with two phorbol esters, phorbol-12,13-didecanoate (PDD) (filled circle), and 12-*O*-tetradecanoylphorbol-13-acetate (TPA) (open circle) for the induction of trypsin-resistant adhesion (*a*) and reduction of SDR cells (*b*) in cultures of C3H-10T1/2 cells. The growth medium of exponentially proliferating cultures that contained 10% fetal bovine serum was replaced with medium in which the serum had been reduced to 0.5%. The indicated concentration of each phorbol ester was present and at 24 hr, cells were counted and expressed in terms of the percentage of cells obtained in control cultures that were treated with an equivalent concentration of acetone solvent alone (± S.E.M., $n = 6$).

tion of trypsin-resistant adhesion, and the spatial relationships of these respective phenomena. Measurement of spatial factors in cell cycle and apoptosis control is a technically difficult task that led us to develop a relatively simple statistical approach which we refer to as *spatial factor analysis*. Both total nuclear density (i.e., unlabeled nuclei) and 24-hour cumulative [^{3}H]dThd-labeled nuclear density were determined in replicate culture plates by scanning between 250 and 850 microscope fields of 3 x 10^{-3} cm^2 each. This information was then analyzed by constructing a frequency distribution and graphically presenting the data in the form of a normal equivalent deviate plot (Finney 1971). It can be shown that the

TABLE 2 COMPARISON OF THE EFFECTS OF PHORBOL ESTERS AND RETINOIDS ON ADHESION MARKERS IN THE C3H-10T1/2 CELLS

	pD_2 [a]			Relative activities	
	SDR [b]	TR [c]	promotion	mitogen	adhesion
TPA	8.72	8.00	++++	++++	++++
4-O-Methyl TPA	7.30	6.30	-	++	+
PDD	10.00	10.40	++++	++++	n.a.
12-O-Tetradecanoyl phorbol	7.22	7.40	-	+	n.a.
Phorbol-12,13-diacetate	<6.00	<6.00	+	-	n.a.
Phorbol-12,13,20-triacetate	<6.00	<6.00	-	-	n.a.
4-(a)OH-phorbol didecanoate	<6.00	<6.00	-	n.a.	n.a.
trans-Retinoic acid	7.25	7.82	antagonist	-	++
Retinyl acetate	n.a.	8.30	antagonist	-	+++

n.a. indicates not assayed.
[a] pD_2 values derived from dose response curves where $pD_2 = -\log(ED_{50})$.
[b] SDR: Response based on number of serum-deprivation-released cells.
[c] TR: Response based on counts of trypsin-resistant cells.

labeled nuclear density (R) distribution can be normalized by the simple power function

$$f(R) = e^{1/q\text{Ln}R}$$

where R is the radiolabeled nuclear density and $q \geq 1.000$. Therefore, when q is 1.000, the response density is normal and the response of any cell would be described by chance alone. However, increased values of q would indicate increased cell cooperativity in G_1/S transition. Therefore, the response of any particular cell becomes a function of both drug concentration and proximity of that cell to other responding cells. Empirically determined values of q are presented in Table 3. Note that TPA increased cell S-phase entry cooperativity as well as the number of responding cells, suggesting that cell:cell interactions are intimately involved in TPA modulation of apoptosis and cell-cycle activation.

Although it is not possible to measure directly the spatial distribution of apoptotic cells in these cultures, we can infer

TABLE 3 *DOSE-RESPONSE RELATIONSHIP OBTAINED FOR TPA AND 4-O-METHYL TPA RELATIVE TO THE RESPONSE DENSITY DISTRIBUTION PARAMETER q[1] IN QUIESCENT CULTURES [2] OF C3H-10T1/2 CELLS AT 350+ CELLS/mm [2]*

Treatment	Labeled nuclei/ mm^2	$m_{0.95}$[3]	Optimal q
Control	9.05	0.71	2.16
TPA 10^{-9} M	10.58	1.20	1.80
10^{-8} M	35.17	2.81	2.00
10^{-7} M	90.27	3.84	3.13
10^{-6} M	67.03	3.21	3.70
4-O-Methyl TPА 10^{-8} M	12.52	0.84	2.27
10^{-7} M	11.53	0.93	1.64
10^{-6} M	78.88	3.28	2.00

[1] Optimal q values were calculated based on $f(R) = e^{1/q\text{Ln}R}$, where R is labeled nuclear density determined from 250–850 scanning fields on duplicate plates. Each field was 3×10^{-3} cm^2 (see text).

[2] All cultures were arrested by 24-hr serum deprivation during exponential growth 4 days after seeding. Drugs were added in 10 μl/5 ml culture volume (control received 10 μl of 50% ethanol solvent only) along with 1 μCi/5 ml [³H]dThd at a specific activity of 45–55 Ci/mmole as described in Fig. 1. Cultures were washed and fixed after 24 hr and prepared for autoradiography directly on the culture dish.

[3] $m_{0.95}$ is the 95% confidence interval for the mean labeled density.

from the data on TPA inhibition and S-phase entry that inhibition of apoptosis exhibits cooperativity among cells. This view is consistent with the qualitative evidence that cells secrete factors that enhance the survival or the proliferation of neighboring cells. Similar observations have been reported consistently for several decades which indicate that cells exposed to cytotoxic stimuli produce diffusible factors that enhance survival (i.e., inhibit death) of other cells, stimuli now known to be capable of inducing apoptosis, such as X irradiation (Revesz 1956), UV irradiation (Schorpp et al. 1984), cytotoxic drugs (Millar et al. 1978), and in vitro passage of primary cells (Lynch et al. 1986). Since both TPA and retinoids have been shown to inhibit gap junction cell:cell communications (Yotti et al. 1979; Mehta et al. 1986), we can speculate that these agents stimulate production and/or release of cooperative growth factors which may be similar to those produced constitutively by apoptotic cells.

The mechanism of trypsin-resistant adhesion remains unknown, as does the relationship between cell cycle and adhesion. Cell adhesion mechanisms, however, have been linked to cell survival and cell-cycle control in association with tumor promoters such as TPA (Shiba and Kanno 1989, 1990; Shibata et al. 1989). Nonetheless, it is clear that changes in cell-cycle control and the inhibition of apoptosis are associated with changes in cell adhesion mechanisms marked by trypsin-induced loss of cell shape and the dependency of adhesion on Ca^{++}. Other agents believed to inhibit apoptosis, such as transforming growth factor (see Gerschenson and Rotello, this volume), are also known to regulate adhesion receptors (Heino and Massague 1989; Heino et al. 1989). If we consider future directions for research, it is worthy to note that the molecular biology of apoptosis is inextricably bound to the molecular biology of cell-cycle control. If early events in apoptosis involve specific changes in chromatin structure of the kind mentioned earlier in this chapter, it should be kept in mind that these processes take place in single specific cells and in cells of interacting populations. Therefore, if a certain cell initiates apoptosis as a consequence of internal damage, for example, to DNA caused by radiation, then that cell also has the capability of modifying the probability that neighboring cells do the same. Contrary to the cooperative interactions leading to S-phase entry, experimental evidence suggests that cell interactions involving apoptotic cells lead to inhibition of apoptosis in neighboring cells. This fact may be intuitively attractive, since all cells in any population may be exposed to the same toxic insult, yet only a portion of the population may die. Such empirical observations are often attributed to purely descriptive effects such as "resistance" or even differing cell-cycle states. In one particular instance, an investigator attributed TPA inhibition of apoptosis to inhibition of "nonadhesion" in an adhesion-dependent cell (pers. comm.); such descriptive physiology is not hypothesis and, therefore, reveals little about the nature of cell growth and death.

Apoptosis and Multistage Carcinogenesis: A Reassessment of the Model

If low-level DNA damage initiates apoptosis in a portion of a cell population, blockade of apoptosis through a mechanism independent of the nature of the original damage should be expected to increase the expression of chromosomal aberration, mutagenesis, and presumably carcinogenesis. A substantial amount of published data is in accord with this view, the interpretation of the respective authors notwithstanding. Consider here the conclusions that phorbol esters act as tumor promoters by virtue of their ability to induce cytogenetic abnormalities. Several investigators reported that TPA increased the expression of chromosomal aberrations (Callen and Ford 1983; Gainer et al. 1984; Dzarlieva-Petrusevska and Fusenig 1985). These observations were in agreement with those of other workers who reported that TPA itself was capable of inducing DNA damage (Birnboim 1982; Hartley et al. 1985; Snyder 1985), which was an attractive point of view based on a strict interpretation of the somatic mutation theory. However, Färber and Kinzel (1990) recently reported that a more detailed examination of the TPA effect revealed that the chromosomal aberration found in TPA-treated HeLa cells is identical in kind to that found to occur spontaneously. These authors go on to conclude that TPA enhanced the expression of a preexisting mechanism of clastogenesis. From this example, it can be seen that TPA blockade of apoptosis would result in the "expression" of cytogenetic aberration, particularly if that aberration reflected an inherent genetic abnormality present in established transformed cell lines in vitro.

If we consider apoptosis as a mechanism for maintenance of genomic fidelity, we are led to speculate that there are endogenous cellular factors capable of blocking this protective mechanism. The consequence presumably would be expression of mutagenic events including frank malignant transformation. Epstein-Barr virus gene products in infected T lymphocytes have been associated with blockade of apoptosis in those cells (Gregory et al. 1991). Dihydroteleocidin-B (DHTB) was shown to block apoptosis in C3H-10T1/2 cells following a variety of toxic insults such as incorporation of very high specific-activity [^{3}H]dThd, or serum growth factor deprivation (Tomei et al. 1988). Using these same cells, Hsiao et al. (1984) reported that DHTB and TPA enhanced oncogene-induced transformation through a mechanism not related to increased transcription, which led these authors to speculate that DHTB and TPA induced expression of new host genes. There is further evidence that oncogenes may actually involve products that block apoptosis. Hockenbery et al. (1990) reported that the product of the bcl-2 oncogene is an inner mitochondrial membrane protein that inhibits apoptosls.

Clouston and Kerr (1985) proposed that immune cells may recognize virus-infected cells and eliminate them through a mechanism of induced apoptosis (see Kerr and Harmon, this volume). However, there is evidence that viruses to some extent may attribute their success as adventitious agents to their ability to integrate with the host cell genes associated with modulation of apoptosis. We have shown that TPA markedly enhanced SV40-mediated immortalization of primary human skin epithelial cells (Tomei and Glaser 1988). However, viruses such as SV40 may be closely integrated with host cell apoptosis genes since UV irradiation of infected cells results in expression of infectious particles (Nomura et al. 1983). Such phenomena are not limited to SV40 since other viruses, most notably human immunodeficiency virus (HIV), can be induced by initiation of apoptosis. Valerie et al. (1988) demonstrated that induction of apoptosis in infected lymphocytes was accompanied by expression of HIV, which suggests that these viruses are closely linked at the gene level with control of apoptosis. In the case of HIV, it is distinctly possible that this virus itself may be involved in modulation of apoptosis in many cell types, including cytotoxic T lymphocytes. In related observations, Kotler et al. (1986) reported accumulation of apoptotic epithelial cells in the rectal crypt region of patients with AIDS and early suppression of gastrointestinal function. It is not clear whether the accumulation of apoptotic cells in tissues of AIDS patients is simply a reflection of the loss of cell-mediated immune function, or whether HIV specifically interferes with control of apoptosis. Since apoptotic cells have been shown to have modified surface macrophage recognition sites, accumulation of apoptotic cells may also reflect either failure to produce these sites on target apoptotic cells, or failure of immune cells to recognize the sites.

CONCLUDING THOUGHTS

It is interesting to speculate on the nature of the hypothetical apoptosis genes in view of our present, albeit rudimentary, knowledge of the process. Two general points can be made: First, the genes are likely to be highly conserved in eukaryotic cells since apoptosis can be expected to comprise a fundamentally critical array of gene-directed events that have been modified to various extents by differentiation. Second, the genes can involve products that either mediate degradative metabolic cascades or mediate positive feedback loops whereby apoptosis is inhibited.

Our laboratory has presented evidence that expression of the human amyloid β protein precursor (AβPP) gene is initiated in basal crypt epithelial cells of the dog ileum by γ irradiation (Tomei et al. 1991). This gene has also been associated with nexin-II, a cell-surface protein having antiprotease activity (Oltersdorf et al. 1989), and with intercellular growth regulation in its secreted form (Saitoh et at. 1989). Recently, this gene has been the subject of substantial attention since reports by Hardy and colleagues (see Goate et al. 1991) that Alzheimer's disease is related to the presence of a mutation in the AβPP gene resulting in abnormal processing of the protein product and accumulation in neuritic plaques. Therefore, it is possible that the AβPP gene product is linked to the secretion of anti-apoptotic protein, which then interacts at the surface sites of neighboring cells. In the presence of an apoptotic signal, the early apoptotic cells would secrete an APP protein, which would then alter surface characteristics of other cells, blocking their initiation of apoptosis and in some instances stimulating replication. This is consistent with observations that the AβPP gene product modulates cell adhesion and is induced by FGF in PC12 nerve cells (Schubert et al. 1989; Yanker et al. 1990). Alzheimer's disease could then be a consequence of a failure in APP-mediated modulation of apoptosis leading to progressive cell loss and accumulation of APP due to lack of feedback inhibition of gene expression. Not all data are consistent with this simple view, however. Amyloid β protein is neurotoxic, and Yanker et at. (1990) have presented evidence that this toxicity is enhanced by nerve growth factor. Therefore, as with other genes, the connection with apoptosis remains tenuous and is the subject of continuing research.

Apoptosis is a concept that has led us to conclude that there may be a triage mechanism in all cells which determines whether cells have sustained damage insufficient to cause catastrophic loss of function but sufficient to increase risk of heritable changes. This mechanism not only accounts for cell deletion, but the deleted cells may also be directly responsible to a significant extent for the proliferation and/or survival of other cells. Ultimately, the process of apoptosis is likely to be a modulated phenomenon having initiators and inhibitors acting both intracellularly and intercellularly. The purpose of such a system is probably maintenance of genetic fidelity, minimization of phenotype variation, and elimination of genotype alteration. Apoptosis is more accurately considered to be not a programmed death, but rather a programmed survival.

ACKNOWLEDGMENTS

I acknowledge the support provided to me by the Arthur G. James Cancer Hospital and Research Institute, The Ohio State

University Comprehensive Cancer Center, and Ross Laboratories of Columbus, Ohio. I acknowledge the collaboration of Drs. Charles Wenner, Peter Kanter, and Fred Cope, and Mr. John Shapiro, whose individual commitments to research and contribution of ideas about apoptosis have made this work so interesting to me.

REFERENCES

Ackermann, R. 1970. *The philosophy of science*. Pegasus, New York.

Alles, A., K. Alley, J.C. Barrett, R. Buttyan, A. Columbano, F.O. Cope, E.A. Copelan, R.C. Duke, P.B. Farel, D. Goldgaber, L.E. Gershenson, D.R. Green, K.V. Honn, J. Hully, J.T. Isaacs, J.F.R. Kerr, P.H. Krammer, R.A. Lockshin, D.P. Martin, D.J. McConkey, J. Michaelson, R. Schulte-Hermann, A.C. Server, B. Szende, L.D. Tomei, T.R. Tritton, S.R. Umansky, K. Valerie, and H.R. Warner. 1991. Apoptosis: A general comment. *FASEB J.* **5:** 2127.

Alnemri, E.S. and G. Litwack. 1990. Activation of internucleosomal DNA cleavage in human CEM lymphocytes by glucocorticoid and novobiocin. *J. Biol. Chem.* **265:** 17323.

Arends, M.J., R.G. Morris, and A.H. Wyllie. 1990. Apoptosis: The role of the endonucleases. *Am. J. Pathol.* **136:** 593.

Birnboim, H.C. 1982. DNA strand breakage in human leukocytes exposed to a tumor promoter, phorbol myristate acetate. *Science* **215:** 1247.

Callen, D.F. and J.H. Ford. 1983. Chromosome abnormalities in chronic lymphocytic leukemia revealed by TPA as a mitogen. *Cancer Genet. Cytogenet.* **10:** 87.

Clouston, W.M. and J.F.R. Kerr. 1985. Apoptosis, lymphocytotoxicity and the containment of viral infections. *Med. Hypotheses* **18:** 399.

Cohen, J.J., R.C. Duke, R. Chervenak, K.S. Selline, and L.K. Olson. 1985. DNA fragmentation in targets of CTL: An example of programmed cell death in the immune system. *Adv. Exp. Med. Biol.* **184:** 439.

Cole, L.J. and M.E. Ellis. 1957. Radiation-induced changes in tissue nucleic acids: Release of soluble deoxynucleotides from the spleen. *Radiat. Res.* **7:** 508.

Duke, R.C., K.S. Sellins, and J.J. Cohen. 1988. Cytotoxic lymphocyte-derived lytic granules do not induce DNA fragmentation in target cells. *J. Immunol.* **141:** 2191.

Duker, N.J., D.M. Hart, and C.L. Grant. 1982. Stability of the DNA apyrimidinic site. *Mutat. Res.* **103:** 101.

Dzarlieva-Petrusevska, R.T. and N.E. Fusenig. 1985. Tumor promoter 12-O-tetradecanoylphorbol-13-acetate (TPA)-induced chromosome aberrations in mouse keratinocyte cell lines: A possible genetic mechanism of tumor promotion. *Carcinogenesis* **6:** 1447.

Elkind, M.M. 1985. DNA damage and cell killing. *Cancer* **56:** 2351.

Färber, B. and V. Kinzel. 1990. Distribution of phorbol ester TPA-induced structural chromosomal aberrations in HeLa cells. *Carcinogenesis* **11:** 2067.

Fesus, L. and V. Thomazy. 1988. Searching for the function of tissue transglutaminase: Its possible involvement in the biochemical pathway of programmed cell death. *Adv. Exp. Med. Biol.* **231:** 119.

Finney, P.J. 1971. *Probit analysis,* 3rd edition. Cambridge University Press, Cambridge.

Fortini, P., M. Bignami, and E. Dogliotti. 1990. Evidence for AP site formation related to DNA-oxygen alkylation in CHO cells treated with ethylating agents. *Mutat. Res.* **236:** 129.

Furtwängler, J.A., S.H. Hall, and L.K. Koskinen-Moffet. 1985. Sutural morphogenesis in the mouse calvaria: The role of apoptosis. *Acta Anat.* **124:** 74.

Gainer, H.S.C., S. Schor, and A.R. Kinsella. 1984. Susceptibility of skin fibroblasts from individuals genetically predisposed to cancer to transformation by the tumour promoter 12-O-tetradecanoyl-phorbol-13-acetate. *Int. J. Cancer* **34:** 349.

Goate, A., M.-C. Chartier-Harlin, M. Mullan, J. Brown, F. Crawford, L. Fidiani, L. Giuffra, A. Haynes, N. Irving, L. James, R. Mant, P. Newton, K. Rooke, P. Roques, C. Talbot, R. Williamson, M. Rossor, M. Owen, and J. Hardy. 1991. Segregation of a missense mutation in the amyloid precursor protein gene with familial Alzheimer's disease. *Nature* **349:** 704.

Gregory, C.D., C. Dive, S. Henderson, C.A. Smith, G.T. Williams, J. Gordon, and A.B. Rickinson. 1991. Activation of Epstein-Barr virus latent genes protects human B cells from death by apoptosis. *Nature* **349:** 612.

Hartley, J.A., N.W. Gibson, L.A. Zwelling, and S.H. Yuspa. 1985. Association of DNA strand breaks with accelerated terminal differentiation in mouse epidermal cells exposed to tumor promoters. *Cancer Res.* **45:** 4864.

Heidelberger, C., A.E. Freeman, R.J. Pienta, A. Sivak, J.S. Bertram, B.C. Casto, V.C Dunkel, M.W. Francis, T. Kakunaga, J.B. Little, and L.M. Schechtman. 1983. Cell transformation by chemical agents: A review of the literature. *Mutat. Res.* **114:** 284.

Heino, J. and J. Massague. 1989. Transforming growth factor-beta switches the pattern of integrins expressed in MG-63 human osteosarcoma cells and causes a selective loss of cell adhesion to laminin. *J. Biol. Chem.* **264:** 21806.

Heino, J., R.A. Ignotz, M.E. Hemler, C. Crouse, and J. Massague. 1989. Regulation of cell adhesion receptors by transforming growth factor-β. Concomitant regulation of integrins that share a common beta 1 subunit. *J. Biol. Chem.* **264:** 380.

Hitchcock, S.E. 1980. Actin-deoxyribonuclease I interaction. *J. Biol.*

Chem. **255:** 5668.

Hockenbery, D., G. Nunez, C. Millman, R.D. Schreiber, and S.J. Korsmeyer. 1990. Bcl-2 is an inner mitochondrial membrane protein that blocks programmed cell death. *Nature* **348:** 334.

Hollander, M.C. and A.J. Fornace. 1989. Induction of *fos* RNA by DNA-damaging agents. *Cancer Res.* **49:** 1687.

Hsiao, W.L., S. Gattoni-Celli, and I.B. Weinstein. 1984. Oncogene-induced transformation of C3H-10T1/2 cells is enhanced by tumor promoters. *Science* **226:** 552.

Ignotz, R.A., J. Heino, and J. Massague. 1989. Regulation of cells adhesion receptors by transforming growth factor-β. *J. Biol. Chem.* **264:** 389.

Kanter, P.M. and H.S. Schwartz. 1980. Post-repair DNA damage in X-irradiated cultured human tumour cells. *Int. J. Radiat. Biol.* **38:** 483.

Kanter, P.M., L.D. Tomei, and C.E. Wenner. 1982. Phorbol ester inhibition of DNA fragmentation. In *Proceedings of the 13th International Cancer Congress*, Seattle, Washington, p. 542.

Kanter, P.M., K.J. Leister, L.D. Tomei, P.A. Wenner, and C.E. Wenner. 1984. Epidermal growth factor and tumor promoters prevent DNA fragmentation by different mechanisms. *Biochem. Biophys. Res. Commun.* **118:** 392.

Kennedy, A.R., G. Murphy, and J.B. Little. 1980. Effect of time and duration of exposure to 12-*O*-tetradecanoylphorbol-13-acetate on X-ray transformation of C3H-10T1/2 cells. *Cancer Res.* **40:** 1915.

Kerr, J.F.R. and J. Searle. 1973. Deletion of cells by apoptosis during castration-induced involution of the rat prostate. *Virchows Arch. Abt. B. Zellpathol.* **13:** 87.

Kerr, J.F.R., A.H. Wyllie, and A.R. Currie. 1972. Apoptosis: A basic biological phenomenon with wide-ranging implication in tissue kinetics. *Br. J. Cancer* **26:** 239.

Kotler, D.P., S.C. Weaver, and J.A. Terzakis. 1986. Ultrastructural features of epithelial cell degeneration in rectal crypts of patients with AIDS. *Am. J. Surg. Pathol.* **10:** 531.

Kurnick, N.B., B.W. Massey, and G. Sandeen. 1959. The effect of radiation on tissue DNA. *Radiat. Res.* **11:** 101.

Kyprianou, N. and J.T. Isaacs. 1988. Activation of programmed cell death in the rat ventral prostate after castration. *Endocrinology* **122:** 552.

Laval, J., S. Boiteux, and T.R. O'Connor. 1990. Physiological properties and repair of apyrimidinic/apurinic sites and imidazole ring-opened guanidines in DNA. *Mutat. Res.* **233:** 73.

Limas, C.J. 1981. Phosphorylation of myocardial nuclei enhances their susceptibility to deoxyribonuclease I. *Biochem. Biophys. Res. Commun.* **100:** 1347.

Lutz, W.K. 1990. Dose-response relationship and low dose extrapolation in chemical carcinogenesis. *Carcinogenesis* **11:** 1243.

Lynch, M.P., S. Nawaz, and L.E. Gerschenson. 1986. Evidence for soluble factors regulating cell death and cell proliferation in primary cultures of rabbit endometrial cells grown on collagen. *Proc. Natl. Acad. Sci.* **83:** 4784.

Malicka-Blaszkiewicz, M. and J.S. Roth. 1981. Some factors affecting the interaction between actin in leukemic L1210 cells and DNase I. *Biochem. Biophys. Res. Commun.* **102:** 594.

Matyasova, J., M. Skalka, and M. Cejkova. 1979. Regular character of chromatin degradation in lymphoid tissues after treatment with biological alkylating agents. *Folia Biol.* **25:** 380.

McCormick, P.J. and J.S. Bertram. 1982. Differential cell cycle phase specificity for neoplastic transformation and mutation to ouabain resistance induced by N-methyl-N'-nitro-N-nitrosoguanidine in synchronized C3H10T1/2 C18 cells. *Proc. Natl. Acad. Sci.* **79:** 4342.

Mehta, P.P., J.S. Bertram, and W.R. Loewenstein. 1986. Growth inhibition of transformed cells correlates with their junctional communication with normal cells. *Cell* **44:** 187.

Millar, J.L., N.M. Blackett, and B.N. Hudspith. 1978. Enhanced post-irradiation recovery of the haemopoietic system in animals pretreated with a variety of cytotoxic agents. *Cell Tissue Kinet.* **11:** 543.

Nikonova, L.V., P.A. Nelipovich, and S.R. Umansky. 1982. The involvement of nucleases in rat thymocyte DNA degradation after γ-irradiation. *Biochim. Biophys. Acta* **699:** 281.

Nomura, S., N. Shobu, and M. Oishi. 1983. Tumor promoter 12-O-tetradecanoylphorbol 13-acetate stimulates simian virus 40 induction by DNA-damaging agents and tumor initiators. *Mol. Cell. Biol.* **3:** 757.

Nover, L. and H. Reinbothe, eds. 1982. Biochemistry of gene expression. In *Cell Differentiation*, p. 23. Springer Verlag, Berlin.

Oltersdorf, T., L.C. Fritz, D.B. Schenk, I. Lieberburg, K.L. Johnson-Wood, E.C. Beattie, P.J. Ward, R.W. Blacher, H.F. Dovey, and S. Sinha. 1989. The secreted form of the Alzheimer's amyloid precursor protein with the Kunitz domain is protease nexin-II. *Nature* **341:** 144.

Promwichit, P., M.G. Sturrock, and I.V. Chapman. 1982. Depressed DNaseI inhibitor activity and delayed DNA damage in X-irradiated thymocytes. *Int. J. Radiat. Biol.* **42:** 565.

Rao, K.M.K. 1985. Phorbol esters and retinoids induce actin polymerization in human leukocytes. *Cancer Lett.* **28:** 253.

Revesz, L. 1956. Effect of tumor cells killed by X-rays upon the growth of admixed viable cells. *Nature* **178:** 1391.

Saitoh, T., M. Sundsmo, J.-M. Roch, N. Kimura, G. Cole, D. Schubert, T. Oltersdorf, and D.B. Schenk. 1989. Secreted form of amyloid β protein precursor is involved in the growth regulation of fibroblasts. *Cell* **58:** 615.

Savill, J., I. Dransfield, N. Hogg, and C. Haslett. 1990. Vitronectin receptor-mediated phagocytosis of cells undergoing apoptosis. *Nature* **343:** 170.

Schorpp, M., U. Mallick, H.J. Ramsdorf, and P. Herrlich. 1984. UV-induced extracellular factor from human fibroblasts communicates the UV response to non-irradiated cells. *Cell* **37:** 861.

Schubert, D., L.W. Lin, T. Saitoh, and G. Cole. 1989. The regulation of amyloid beta protein precursor secretion and its modulatory role in cell adhesion. *Neuron* **3:** 689.

Shiba, Y. and Y. Kanno. 1989. Modulation of survival and proliferation of BSC-1 cells through changes in spreading behavior caused by the tumor promoting phorbol ester TPA. *Cell Struct. Funct.* **14:** 685.

———. 1990. Survival of BSC-1 cells through the maintenance of cell volume brought about by epidermal growth factor depends on attachment to the substratum. *Experientia* **46:** 492.

Shiba, Y., Y. Sasaki, Y. Kanno, and F. Grinnell. 1989. Enhanced binding of fibronectin-coated latex beads to quiescent 3T3-L1 cells is correlated with escape from growth arrest. *Exp. Cell Res.* **182:** 144.

Sklar, M.D. 1988. The *ras* oncogenes increase the intrinsic resistance of NIH 3T3 cells to ionizing radiation. *Science* **239:** 645.

Snyder, R.D. 1985. An examination of the DNA damaging and repair inhibitory capacity of phorbol myristate acetate in human diploid fibroblasts. *Carcinogenesis* **6:** 1667.

Sullivan, N.F. and A.E. Willis. 1989. Elevation of c-*myc* protein by DNA strand breakage. *Oncogene* **4:** 1497.

Swingle, K.F. and L.J. Cole. 1967. Radiation-induced free polydeoxyribonucleotides in lymphoid tissues: A product of the action of neutral deoxyribonuclease (DNase 1). *Radiat. Res.* **30:** 81.

Tomei, L.D. and R. Glaser. 1988. Increased sensitivity of primary human epithelial cells to SV40 immortalization following phorbol ester dependent Epstein-Barr virus transformation. In *Epstein-Barr virus and human disease, II* (ed. D.V. Ablashi et al.) p. 495. Humana Press, Clifton, New Jersey.

Tomei, L.D. and C.E. Wenner. 1981. Phenotypic expression of malignant transformation and its relationship to energy metabolism. In *The transformed cell* (ed. I.L. Cameron and T.B. Pool), p. 163. Academic Press, New York.

Tomei, L.D., J.C. Cheney, and C.E. Wenner. 1981. The effect of phorbol esters on the proliferation of C3H-10T1/2 mouse fibroblasts: Consideration of both stimulatory and inhibitory effects. *J. Cell. Physiol.* **107:** 385.

Tomei, L.D., P.M. Kanter, and C.E. Wenner. 1985. Tumor promoters inhibit radiation-induced lethal DNA fragmentation in mouse embryonic fibroblasts. *J. Cell. Biochem.* **9C:** 23.

———. 1988. Inhibition of radiation-induced apoptosis in vitro by tumor promoters. *Biochem. Biophys. Res. Commun.* **155**: 324.

Tomei, L.D., P.M. Kanter, K.J. Leister, and C.E. Wenner. 1986. Do tumor promoters inhibit apoptosis? In *Proceedings of the 14th International Cancer Congress*, Budapest, Hungary, p. 449.

Tomei, L.D., O.G. Issinger, J. Shapiro, A.H. McArdle, and F.O. Cope. 1991. Induction of the expression of the Alzheimer's gene by radiation in intestinal epithelial cells: Implications for a novel paradigm in gene-directed cell death. *FASEB J.* **5**: 7171. (Abstr.)

Umansky, S.R., B.A. Korol, and P.A. Nelipovich. 1981. In vivo DNA degradation in thymocytes of γ-irradiated or hydrocortisone-treated rats. *Biochim. Biophys. Acta* **655**: 9.

Utsumi, H., M.L. Shibuya, T. Kosaka, W.E. Buddenbaum, and M.M. Elkind. 1990. Abrogation by novobiocin of cytotoxicity due to the topoisomerase II inhibitor amsacrine in Chinese hamster cells. *Cancer Res.* **50**: 2577.

Valerie, K., A. Delers, C. Bruck, C. Thiriart, H. Rosenberg, C. Debouck, and M. Rosenberg. 1988. Activation of human immunodeficiency virus type 1 by DNA damage in human cells. *Nature* **333**: 78.

Wenner, C.E., L.D. Tomei, and K. Leister. 1984. Tumor promoters: An overview of membrane-associated alterations and intracellular events. *Transplant. Proc.* **16**: 381.

Williams, J.R., J.B. Little, and W.U. Shipley. 1974. Association of mammalian cell death with a specific endonucleolytic degradation of DNA. *Nature* **252**: 754.

Williamson, R. 1970. Properties of rapidly labelled deoxyribonucleic acid fragments isolated from the cytoplasm of primary cultures of embryonic mouse liver cells. *J. Mol. Biol.* **51**: 157.

Wyllie, A.H. 1980. Glucocorticoid-induced thymocyte apoptosis is associated with endogenous endonuclease activation. *Nature* **284**: 555.

Wyllie, A.H., R.G. Morris, A.L. Smith, and D. Dunlop. 1984. Chromatin cleavage in apoptosis: Association with condensed chromatin morphology and dependence on macromolecular synthesis. *J. Pathol.* **142**: 67.

Yanker, B.A., A. Caceres, and L.K. Duffy. 1990. Nerve growth factor potentiates the neurotoxicity of beta amyloid. *Proc. Natl. Acad. Sci.* **87**: 9020.

Yotti, L.P., C.C. Chang, and J.E. Trosko. 1979. Elimination of metabolic cooperation in Chinese hamster cells by a tumor promoter. *Science* **206**: 1089.

Index

Italicized page numbers refer to figures or tables.